Top-notch Academic Programs Project of Jiangsu Higher Education Institutions(PPZY2015A046)
A Project Funded by the Priority Academic Program Development of Jiangsu Higher Education Institutions
National Key Research and development program(2016YFC0600901)
National Natural Science Foundation of China(51574224,51604268)
New Engineering Research and Practice Project of Ministry of Education,China

Rock Excavation and Support

Li Guichen   Elmo Davide   Yan Shuai

China University of Mining and Technology Press

内容简介

岩石开挖对煤矿开采具有十分重要的意义。本书详细论述了岩石开挖和支护方法,从岩石强度、岩石数据采集、岩土工程设计、岩土工程数值模拟等方面探讨和解决采矿过程中的岩土工程问题,并给出实例。全书内容丰富,层次清晰,深入浅出,论述有据,图文并茂,具前瞻性、先进性和适用性。

本教材可为采矿工程及相关专业读者提供参考。

图书在版编目(CIP)数据

岩石开挖与支护/李桂臣,(加)大卫·艾蒙(Elmo Davide),闫帅主编. —徐州:中国矿业大学出版社,2018.12
书名原文:Rock Excavation and Support
ISBN 978 - 7 - 5646 - 4317 - 1

Ⅰ. ①岩… Ⅱ. ①李… ②大… ③闫… Ⅲ. ①矿山开采—基础开挖②矿山开采—矿山支护 Ⅳ. ①TD8

中国版本图书馆 CIP 数据核字(2018)第 297526 号

书　　名	岩石开挖与支护
主　　编	李桂臣　[加]大卫·艾蒙(Elmo Davide)　闫　帅
责任编辑	王美柱
出版发行	中国矿业大学出版社有限责任公司
	(江苏省徐州市解放南路　邮编 221008)
营销热线	(0516)83884103　83885105
出版服务	(0516)83995789　83884920
网　　址	http://www.cumtp.com　E-mail:cumtpvip@cumtp.com
印　　刷	江苏淮阴新华印务有限公司
开　　本	787×1092　1/16　印张 11.25　字数 295 千字
版次印次	2018 年 12 月第 1 版　2018 年 12 月第 1 次印刷
定　　价	26.80 元

(图书出现印装质量问题,本社负责调换)

Preface

Geotechnical Engineering is an applied subject and with strong practicality. It is widely used in mining, construction, underground engineering, transportation and railway construction, and plays an increasingly important role in all walks of life.

Rock excavation and support is the main line of this book, focusing on the rock excavation methods and support methods after excavation. The book mainly introduces: rock engineering design, rock strength definition, classification system and failure criteria of rock failure; rock excavation methods by drilling and blasting and mechanical methods and rock blasting damage mechanics; rock bolt, anchor cable support and shotcrete support methods and system support design after excavation; site rock Soil geotechnical engineering case studies and numerical simulation are briefly introduced.

The book is composed of seven chapters, with Elmo Davide, Li Guichen and Yan Shuai as its chief editor. The compilers of each chapter are as follows: the first, second and seventh chapters are Elmo Davide; the third and fourth chapters are Li Guichen; the fifth and sixth chapters are Yan Shuai; in addition, graduate student Wang Xi also participates in the data collation of the book.

In the process of compiling this book, many teachers and experts have cared a lot and supported it and put forward many valuable suggestions. Here I would like to thank them.

In the process of compiling this book, we have consulted many references, which may not be listed in all the references after the article. Thank you to all the authors of the references.

Due to the limitation of the editor's level, there are unavoidable shortcomings or mistakes in the book. Readers are encouraged to send their comments, corrections, criticisms and suggestions to us.

Editor
November 2018

Contents

Chapter 1 Rock mass strength 1
  1.1 Rockmass classification systems 1
  1.2 Mohr-Coulomb failure criterion 11
  1.3 Strength anisotropy and size 15
  1.4 Generalised Hoek Brown failure criterion 17

Chapter 2 Rock engineering design 19
  2.1 Introduction 20
  2.2 Data collection, data uncertainty and data variability 21
  2.3 Factor of safety and probability of failure 33

Chapter 3 Rock excavation methods 42
  3.1 Introduction 42
  3.2 Drill & blast 44
  3.3 Blasting damage in rock 64
  3.4 Mechanical excavation in rock 68

Chapter 4 Rock excavation and support 76
  4.1 Bolts, dowels and cables 76
  4.2 Shotcrete 84
  4.3 Support design 85

Chapter 5 Rock engineering design 94
  5.1 Structurally-controlled instability 95
  5.2 Stress analysis & underground openings 111

Chapter 6 Rock mechanics design (underground) 121
  6.1 Pillar supported mining methods 121
  6.2 Longwall mining 134
  6.3 Stope stability & the stability graph method 140

Chapter 7 Numerical analysis ··· 145
  7.1 Introduction to numerical modelling ································ 145
  7.2 Applications of numerical models to coal mining ················ 151
  7.3 Discrete fracture network modelling ································ 155

References ·· 163

Appendix to section 3.1 RMR tables ·· 171

Chapter 1　Rock mass strength

Reliable estimates of the strength and deformation characteristics of rock masses are required for almost any form of analysis used for the design of slopes, foundations and underground excavations. Different rock mass behavior is shown in Figure 1-1[1,2].

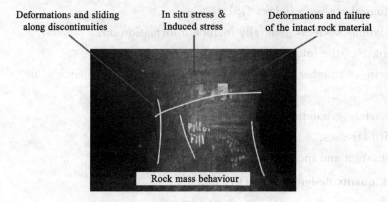

Figure 1-1　Rock mass behaviour

1.1　Rockmass classification systems

Rock specimens are limited in size and therefore represent a very small and highly selective sample of the rock mass from which they were removed (Hoek, 1997)[3,4].

How do we relate the strength and deformability of intact rock samples to the strength and deformability of the in situ rock mass? Figure 1-2 shows the field rock samples.

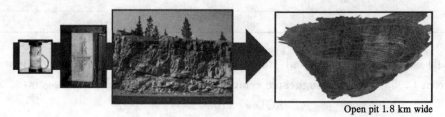

Figure 1-2　Rock sample

Rock mass classification systems represent an attempt to provide guidance on the properties of rock masses.

These classifications were developed primarily for the estimation of the support requirements in tunnels but their use has been expanded to cover many other fields[5,6].

Attempt to combine index properties of intact rocks with the characteristics of field discontinuities.

Use the geological structure and the conditions of the discontinuities to characterise rock mass[7].

Numerous systems proposed but three main ones in use:

① Rock mass rating, RMR.

② Q index.

③ Geological strength index, GSI.

Classifications systems generally include information on:

① Strength of the intact rock material.

② Spacing, number of joint sets and surface properties of the structural discontinuities.

③ Subsurface groundwater.

④ In situ stresses.

⑤ Orientation and inclination of dominant discontinuities.

1.1.1 Rock quality designation (RQD) index

(1) The rock quality designation index (RQD) was developed by Deere et al. (1967) to provide a quantitative estimate of rock mass quality from drill core logs. RQD is defined as the percentage of intact core pieces longer than 100 mm in the total length of core[8,9]. Grading standards are shown in Table 1-1. Figure 1-3 is an example of calculation.

Table 1-1　　　　　　　　　　Core grading table

$RQD/\%$	Drill core quality
<25	Very poor
25~50	Poor
50~75	Fair
75~90	Good
>90	Very good

RQD may also be estimated from the number of discontinuities per unit volume Palmström (2005)[10]. The suggested relationship for clay-free rock masses is:

$$RQD = 110 - 2.5J_v$$

Estimation of discontinuous number per unit volume are shown in Figure 1-4.

Where J_v is the sum of the number of joints per unit length for all joint (discontinuity) sets known as the volumetric joint count[11].

RQD is a directionally dependent parameter and its value may change significantly,

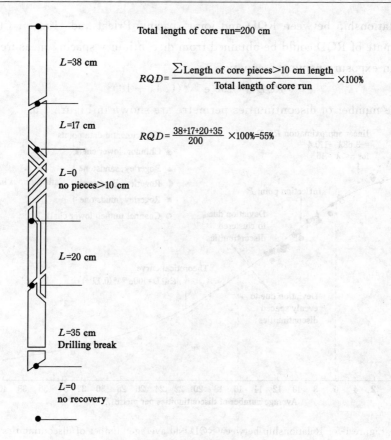

Figure 1-3 Example

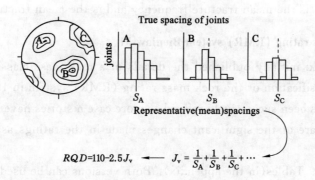

Figure 1-4 The spacing of joints

depending upon the borehole orientation. Does not account for discontinuity orientation, infill, weathering et al.

The use of the volumetric joint count can be quite useful in reducing this directional dependence.

RQD is intended to represent the rock mass quality in situ. Care must be taken to ensure that fractures, which have been caused by handling or the drilling process, are identified and ignored when determining the value of RQD[12,13].

(2) Relationship between RQD and joint spacing. Priest and Hudson (1976) found that an estimate of RQD could be obtained from discontinuity spacing measurements made on core or an exposure using the equation:

$$RQD = 100\ e^{-0.1\lambda}(0.1\lambda + 10)$$

Average number of discontinuities permetre are shown in Figure 1-5.

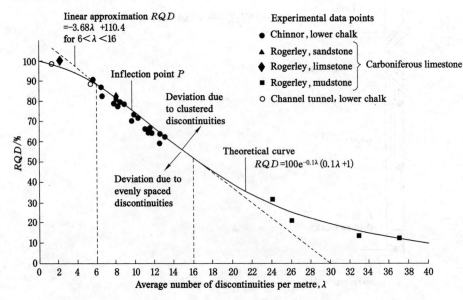

Figure 1-5 Relationship between RQD and average number of discontinuities

Where $\lambda = \dfrac{1}{x}$ is the mean fracture frequency and is the mean fracture spacing.

1.1.2 Rock mass rating (RMR) system-Bieniawski

(1) Bieniawski initially published the details of a rock mass classification called the geomechanics classification or the rock mass rating (RMR) system in 1976.

The RMR has been successively refined as more case histories have been examined and you should be aware of the significant changes made in the ratings assigned to different parameters[14].

1976 and 1989 Tables(in the appendix). Both versions can be used for estimating the strength of rock masses and both versions use the following six parameters (though the ratings assigned to each parameter may be different):

① Uniaxial compressive strength of rock material.
② Rock quality designation (RQD).
③ Spacing of discontinuities.
④ Condition of discontinuities.
⑤ Groundwater conditions.
⑥ Orientation of discontinuities.

In applying RMR classification system, the rock mass is divided into a number of structural regions and each region is classified separately[15,16].

The boundaries of the structural regions usually coincide with a major structural feature such as a fault or with a change in rock type.

In some cases, significant changes in discontinuity spacing or characteristics, within the same rock type, may necessitate the division of the rock mass into a number of small structural regions. Figure 1-6 shows rock tectonic zoning[17]. Table 1-2 shows RMR grading standard. Table 1-3 shows the different between 1976 and 1989.

Figure 1-6 Division of rock structural regions

Table 1-2 RMR rock classification

Rating	100～81	80～61	60～41	40～21	<21
Class No.	I	II	III	IV	V
Description	Very good rock	Good rock	Fair rock	Poor rock	Very poor rock

Table 1-3 RMR ratings 1976 and 1989 versions

Parameter	Rating 1976	Rating 1989
Uniaxial compressive strength of rock material (point load or uniaxial compressive strength)	0 to 15	0 to 15
Rock quality designation (RQD)	3 to 20	3 to 20
Spacing of discontinuities	5 to 30	5 to 20
Condition of discontinuities RMR 1976: General description for condition of discontinuities (no joint length) RMR 1989: General description and additional table with sub-ratings for specific joint conditions (including joint length)	0 to 25	0 to 30
Groundwater conditions	0 to 10	0 to 15
Adjustments for orientation of discontinuities	60 to 0	50 to 0

(2) RMR and support guidelines:

10 m span horseshoe shaped tunnel.

Rock mass subjected to a vertical stress <25 MPa (depth below surface of <900 m).

This table has not had a major revision since 1973.

In many mining and civil engineering applications, steel fibre reinforced shotcrete may be considered in place of wire mesh and shotcrete. Chart of RMR and support guidelines are shown in Figure 1-4.

(3) Modifications to RMR for mining applications:

The RMR system was originally based upon case histories drawn from civil engineering.

Several modifications have been proposed in order to make the classification more relevant to mining applications.

MRMR (modified rock mass rating system for mining) by Laubscher (1977, 1984), Laubscher and Taylor (1976), Laubscher and Page (1990):

The MRMR takes the basic RMR value and adjusts it to account for in situ and induced stresses, stress changes and the effects of blasting and weathering.

A set of support recommendations is associated with the resulting MRMR value.

Laubscher's MRMR system is typically used for caving operations (it was originally derived based upon case histories of block caving in asbestos mines in Africa).

SMR (slope mass rating) classification, introduced by Romana (1985) to specifically. As shown in Figure 1-7.

Relate the RMR classification to slopes, by recognising that joints are an important governing factor for the stability of rock slopes.

1.1.3 Q index

(1) Barton et al. (1974) of the norwegian geotechnical institute (NGI) proposed a tunnelling quality index (Q) for the determination of rock mass characteristics and tunnel support requirements[18].

The numerical value of the index Q varies on a logarithmic scale from 0.001 to a maximum of 1 000 and is defined by:

$$Q = \frac{RQD}{J_n} \times \frac{J_r}{J_a} \times \frac{J_w}{SRF}$$

RQD=Rock quality designation;

J_n=Joint set number;

J_r=Joint roughness number;

J_a=Joint alteration number;

J_w=Joint water reduction number;

SRF=Stress reduction factor.

For specific applications (e.g. stope stability analysis):

Chapter 1 Rock mass strength

Rock mass class	Excavation	Rock bolts (20 mm diameter, fully grouted)	Shotcrete	Steel sets
Ⅰ—Very good rock RMR:81~100	Full face, 3 m advance	Generally no support required except spot bolting		
Ⅱ—Good rock RMR:61~80	Full face, 1~1.5 m advance. Complete support 20 m from face	Locally, bolts in crown 3 m long, spaced 2.5 m with occasional wire mesh	50 mm in crown where required	None
Ⅲ—Fair rock RMR:41~60	Top heading and bench 1.5~3 m advance in top heading. Commence support after each blast. Complete support 10 m from face	Systematic bolts 4 m long, spaced 1.5~2 m in crown and walls with wire mesh in crown	50~100 mm in crown and 30 mm in sides	None
Ⅳ—Poor rock RMR:21~40	Top heading and bench 1.0~1.5 m advance in top heading. Install support concurrently with excavation, 10 m from face	Systematic bolts 4~5 m long, spaced 1~1.5 m in crown and walls with wire mesh	100~150 mm in crown and 100 mm in sides	Light to medium ribs spaced 1.5 m where required
Ⅴ—Very poor rock RMR:<20	Multiple drifts 0.5~1.5 m advance in top heading. Install support concurrently with excavation. Shotcrete as soon as possible after blasting	Systematic bolts 5~6 m long, spaced 1~1.5 m in crown and walls with wire mesh. Bolt invert	150~200 mm in crown and 150 mm in sides, and 50 mm on face	Medium to heavy ribs spaced 0.75 m with steel lagging and forepoling if required. Close invert

Figure 1-7 RMR and support guidelines

$$Q' = \frac{RQD}{J_n} \times \frac{J_r}{J_a}$$

RQD/J_n represents structure of rock mass (block size). Range 100/0.5 to 10/20.

J_r/J_a represents roughness and frictional properties of joint wall and infill (inter block shear strength).

① Rock wall contact.

② Rock wall contact before 10 cm shear.

③ No rock wall contact when sheared.

J_w/SRF represents the "active stresses".

① Loosening load where excavation through shear zones and clay bearing rock.

② Rock stress in competent rock.

③ Squeezing loads in plastic rocks.

(2) Q index (Barton, 2002):

① The first quotient (RQD/J_n), representing the structure of the rock mass, is a crude measure of the block or particle size.

Use the estimated RQD value and the J_n parameter corresponding to the description below. J_n are display in Table 1-4.

Table 1-4 **Joint set number J_n**

Joint set number	J_n
Massive, no or few joints	0.5~1
One joint set	2
One joint set plus random joints	3
Two joint sets	4
Two joint sets plus random joints	6
There joint sets	9
There joint sets plus random joints	12
Four and more joints sets, random, heavily jointed, "sugar-cube", et al	15
Crushed rock, earthlike	20

② The quotient J_r/J_a represents the roughness and frictional characteristics of the joint walls or filling materials.

The ratio will be higher for rough, unaltered joints in direct contact.

When rock joints have thin clay mineral coatings and fillings, the strength is reduced significantly.

Where no rock wall contact exists, the conditions are extremely unfavourable to tunnel stability.

Presence of clay minerals is a deteriorating factor in terms of stability.

Example:

Smooth planar ($J_r=1$) and slightly altered joint walls ($J_a=2$)→$J_r/J_a=0.5$.

③ The parameter J_w is a measure of water pressure.

Water pressure has an adverse effect on the shear strength of joints.

Water may, in addition, cause softening and possible outwash in the case of clay-filled joints.

SRF is a measure of:

Loosening load in the case of an excavation through shear zones and clay bearing rock.

Rock stress in competent rock.

Squeezing loads in plastic incompetent rocks.

(3) Correlations between Q index and RMR:

Proposed correlations between Q index and RMR (Choquet and Hadjigeorgiou, 1993)[21].

These correlations ignore the fact that the two systems are not truly equivalent.

RMR does not consider the stress condition of the rock mass.

The Q index does not consider joint orientation.

The correlations in the table below do not differentiate between RMR_{76} and RMR_{89}.

Example:
$$Q = 10 \text{ and using } RMR = 9\ln Q + 44$$
$$RMR = 65$$

Do not quote RMR values using decimal figures. Proposed correlations between Q index and RMR is display in Table 1-5.

Table 1-5   **Proposed correlations between Q index and RMR**

Correlation	Source	Comments
$RMR = 13.5\log Q + 43$	New Zealand	Tunnels
$RMR = 9\ln Q + 44$	Diverse origin	Tunnels
$RMR = 12.5\log Q + 55.2$	Spain	Tunnels
$RMR = 5\ln Q + 60.8$	S. Africa	Tunnels
$RMR = 43.89 - 9.19\ln Q$	Spain	Mining soft rock
$RMR = 10.5\ln Q + 41.8$	Spain	Mining soft rock
$RMR = 12.11\log Q + 50.81$	Canada	Mining hard rock
$RMR = 8.7\ln Q + 38$	Canada	Tunnels sedimentary rock
$RMR = 10\ln Q + 39$	Canada	Mining hard rock

(4) Q index & support guidelines:

Support design by using the equivalent dimension D_e and ESR. (the latter related to use of excavation and to extent of to which some instability is acceptable)[22,23]

The length L of rock bolts can be estimated from the excavation width B and the excavation support ratio ESR:

$$L = 2 + \frac{0.15B}{ESR}$$

The maximum unsupported span can be estimated from:

$$\text{Maximum span (supported)} = 2 \times ESR \times Q^{0.4}$$

Grimstad and Barton (1993) suggested that the relationship between the value of Q and the permanent roof support pressure P_{roof} can be estimated from:

$$D_e = \frac{\text{Excavation span (diameter or height) (m)}}{\text{Excavation support ratio}(ESR)}$$

$$P_{\text{roof}} = \frac{2\sqrt{J_n} \times Q^{\frac{1}{3}}}{3 J_r}$$

The rock class is display in Figure 1-8.

(5) Using RMR and Q index:

Two approaches can be taken:

① Evaluate the rock mass specifically for the parameters included in the classification methods.

② Accurately characterise the rock mass and then attribute parameter ratings at a later time.

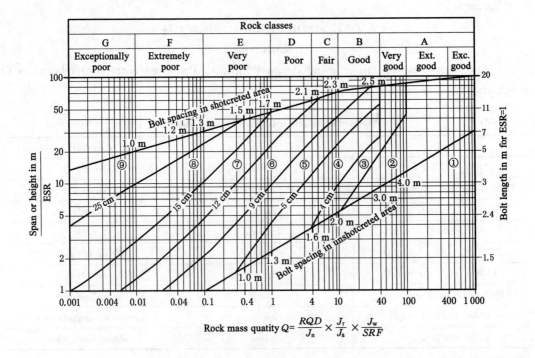

Figure 1-8 Chart of rock classes
①—Unsupported;②—Spot bolting;③—Systematic bolting;④—Systematic bolting
(and unreinforced shotcrete,4~10 cm);
⑤—Fibre reinforced shotcrete and bolting,5~9 cm;⑥—Fibre reinforced shotcrete and bolting,9~12 cm;
⑦—Fibre reinforced shotcrete and bolting,12~15 cm;⑧—Fibre reinforced shotcrete,>15 cm;
reinforced ribs of shotcrete and bolting;⑨—Cast concrete lining

The latter method gives a full and complete description of the rock mass which can easily be translated into various classification index.

If rating values alone had been recorded during mapping, it would be almost impossible to carry out verification studies.

Example (Field notes by Dr. N. Barton): Histograms showing the variations in RQD, J_n, J_r and J_a, as mapped along an exploration drive.

1.1.4 The geological strength index (GSI)

Initially introduced by Hoek (1994) and Hoek et al. (1995), the GSI was developed to overcome some of the deficiencies that had been identified in using the RMR scheme with the Hoek-Brown rock strength criterion[26].

The GSI was developed specifically as a method of accounting for the properties of a discontinuous or jointed rock mass which influence its strength and deformability.

GSI provides a number that combined with the intact rock properties can be used for estimating the reduction in rock mass strength for different geological conditions.

① Blockiness and degree or interlocking.

② Condition of the discontinuity surfaces.

GSI does not explicitly include an evaluation of the uniaxial compressive strength of the intact rock pieces and avoids the double allowance for discontinuity spacing as occurs in the RMR system.

GSI does not include allowances for water or stress conditions which are accounted for in the stress and stability analyses (Hoek-Brown criterion).

1.2 Mohr-Coulomb failure criterion

1.2.1 Intact rock failure criteria

A strength criterion is a relation between stress components which permits the strength developed under various stress combinations to be predicted.

Peak strength criterion.

Residual strength criterion.

Yield (permanent deformation) criterion.

Quantitatively we can express the failure criteria by using the different rock strength (tensile, shear and compressive):

Tensile failure will occur if we have: $\sigma < \sigma_t$;

Compressive failure will occur if we have: $\sigma > \sigma_{ci}$.

Shear failure will involve both normal and shear stresses → shear strength criterion. Zones of overstress and relaxation are display in Figure 1-9.

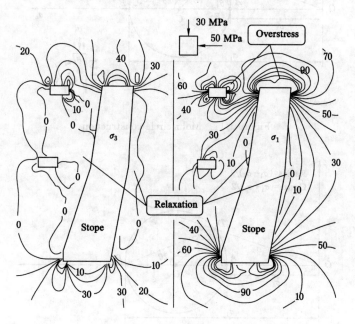

Figure 1-9　Zones of overstress and relaxation

1.2.2 Mohr-Coulomb failure criterion

Shear strength consists of two components: cohesive and frictional.

① C and f are measures of shear strength.

② The higher the values, the higher the shear strength.

Stress diagram can show in Figure 1-10. Figure 1-11 to Figure 1-14 show the Mohr Coulomb failure criterion.

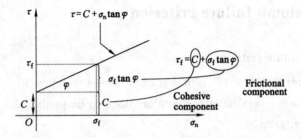

Figure 1-10 Stress diagram

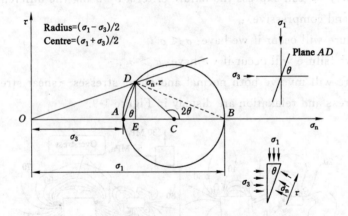

Figure 1-11 Mohr circle construction

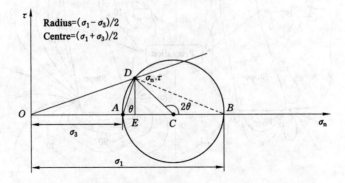

Figure 1-12 Mohr circle construction

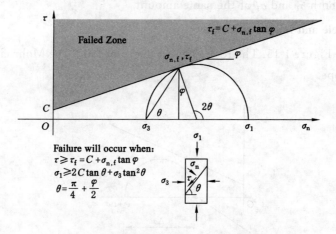

Figure 1-13 Mohr-Coulomb criterion

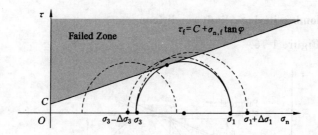

Figure 1-14 Failure zone diagram of rock mass

The normal stress σ_n and the shear stress τ on any plane that has an angle of θ from the maximum principle stress σ_1 direction can be related to the maximum and minimum stress by the following equations:

In terms of major and minor principal stresses[27,28]:

These relationships can also be expressed graphically→Mohr's circles.

$$\sigma_n = \frac{1}{2}(\sigma_1 + \sigma_3) + \frac{1}{2}(\sigma_1 - \sigma_3) \times \cos 2\theta$$

$$\tau = \frac{1}{2}(\sigma_1 - \sigma_3) \times \sin 2\theta$$

$$\sigma_1 = \frac{2C \times \cos \varphi}{1 - \sin \varphi} + \frac{1 + \sin \varphi}{1 - \sin \varphi} \times \sigma_3$$

$$\sigma_n = OA + OE = \sigma_3 + AD \times \cos \theta = \sigma_3 + (\sigma_1 - \sigma_3) \times \cos^2 \theta$$

$$\tau = DE = DC \times \sin 2\theta = \frac{\sigma_1 - \sigma_3}{2} \times \sin 2\theta$$

There are three ways for the circle to "cross" the straight line to reach failure:

① Increase σ_1;

② Decrease σ_3;

③ Decrease both σ_1 and σ_3 of the same amount.

1.2.3 Mohr circle and failure envelope

As shown in Figure 1-15. The rock element does not fail if the Mohr circle is contained within the envelope.

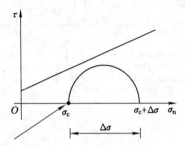

Figure 1-15 Mohr circle and failure envelope

1.2.4 Mohr-Coulomb criterion: limit conditions

As shown in Figure 1-16.

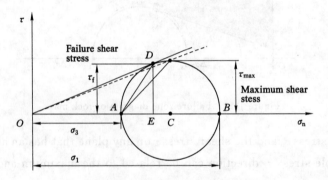

Figure 1-16 Limit conditions

The failure plane is not the plane subjected to the maximum shear stress.

The criterion for failure is maximum obliquity ($\alpha \to \varphi$).

Since it would not be reasonable to admit a frictional resistance in the presence of a tensile normal stress, the Mohr-Coulomb loses its physical validity when the value of σ crosses into the tensile region.

The minimum principal stress σ_3 may be tensile as long as σ_1 remains compressive.

Other theories of failure (e.g. Griffith theory) are more exact in the tensile region.

The Mohr-Coulomb[29,30] theory could be retained by extrapolating the Mohr-Coulomb line into the tensile region up to a point where σ_3 becomes equal to σ_t (tensile strength).

This is equivalent to considering a "tension cut-off". As shown in Figure 1-17.

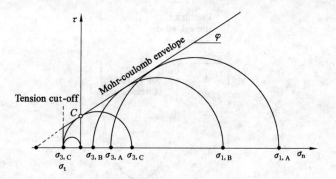

Figure 1-17 Tensile region

1.3 Strength anisotropy and size

So far we have assumed that the mechanical response of rock material is isotropic.

Because of some preferred orientation of the rock fabric, or the presence of bedding/cleavage planes, the behaviour of the rock material will be anisotropic.

The behaviour of transversely isotropic rocks in triaxial compression will vary with the orientation of the plane of isotropy (e.g. foliation plane or bedding plane), with respect to the principal stress directions.

1.3.1 Strength of anisotropic rock material

As shown in Figure 1-18 and Figure 1-19.

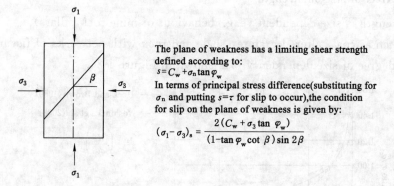

The plane of weakness has a limiting shear strength defined according to:
$s = C_w + \sigma_n \tan \varphi_w$
In terms of principal stress difference (substituting for σ_n and putting $s=\tau$ for slip to occur), the condition for slip on the plane of weakness is given by:
$$(\sigma_1 - \sigma_3)_s = \frac{2(C_w + \sigma_3 \tan \varphi_w)}{(1 - \tan \varphi_w \cot \beta) \sin 2\beta}$$

Figure 1-18 Strength analysis of rock mass

Mechanical behaviour of transversely isotropic rocks in triaxial compression:

The minimum strength occurs when:

$$\tan 2\beta = -\cot \varphi_w \rightarrow \beta = \frac{\pi}{4} + \frac{\varphi_w}{2}$$

1.3.2 Influence of principal stress ratio on failure

Stress ratio k:

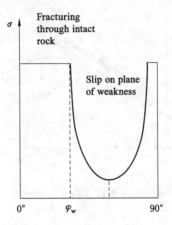

Figure 1-19 Fracturing through intact rock

$$k = \frac{\text{horizontal stress}}{\text{vertical stress}}$$

In a standard triaxial test, the rock is initially subjected to ahydrostatic stress, i. e.:

$$k = \frac{\text{horizontal stress}}{\text{vertical stress}} = \frac{\sigma_3}{\sigma_1} = 1$$

As σ_1 is increased, the value of k is reduced until cracking occurs and eventually peak strength is reached.

This may not be a realistic loading condition and it may be desirable to consider rock behaviour when the stress ratio k is fixed at some value.

1.3.3 The effect of size on strength

Rock strength is size dependent (e. g. behaviour of mine rock pillars).

Rocks such as coal, altered granite, shale and rock with networks of fissures exhibit the greatest degree of size dependency. As shown in Figure 1-20.

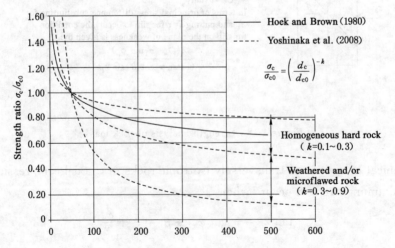

Figure 1-20 Equivalent diameter

Scale effect relations for intact rock UCS proposed by Yoshinaka et al., (2008)-dashed lines. The relation of Hoek and Brown (1980) is also shown for comparison (modified from Pierce et al., 2009).

1.4 Generalised Hoek Brown failure criterion

1.4.1 Introduction: empirical failure criteria

Consider the Mohr's circles below.

We could define an empirical failure criterion represented by the envelope to those Mohr's circles.

The classic strength theories used for other engineering materials are not applicable to rock over a wide range of applied compressive stress conditions, therefore a number of empirical strength criteria have been introduced over the years for practical use.

These criteria usually take the form of a power law in recognition of the fact that peak σ_1 vs. σ_3 and τ vs. σ_n envelopes for rock material are generally concave downwards.

In order to ensure that the parameters used in the power laws are dimensionless, these criteria are best written in normalised form with all stress components being divided by the uniaxial compressive strength of the rock.

Examples: Bieniawski (1974) and Hoek-Brown (1980).

1.4.2 Hoek-Brown criterion

Hoek-Brown failure criterion:

Firstly introduced in 1980, the Hoek-Brown criterion has been since modified several times, most recently by Hoek and Brown (1997) and Hoek et al. (2002).

In section 7 we have introduced the GSI classification system and discussed how GSI is based upon an assessment of the interlocking of rock blocks and the condition of the surfaces between these blocks.

The Hoek-Brown criterion requires GSI to calculate several parameters that are used in the Hoek-Brown equation.

The Hoek-Brown criterion starts from the properties of intact rock and then introduces factors to reduce these properties on the basis of the characteristics of joints in a rock mass (i.e. by using GSI).

Initially, the Hoek-Brown criterion was linked to the RMR_{76} system, it was later modified with the introduction of the geological strength index (GSI).

1.4.3 Generalised Hoek-Brown failure criterion

The generalised Hoek-Brown failure criterion for jointed rock masses is defined by:

$$\sigma_1 = \sigma_3 + \sigma_{ci}\left\{m_b \times \frac{\sigma_3}{\sigma_{ci}} + s\right\}$$

Where σ_1 and σ_3 are the maximum and minimum effective principal stresses at failure,

m_b is the value of the Hoek-Brown constant m_i for the rock material, s and a are constants which depend upon the rock mass characteristics, and σ_{ci} is the uniaxial compressive strength of the intact rock pieces.

In order to use the Hoek-Brown criterion three properties have to be estimated:

① Uniaxial compressive strength σ_{ci} of the intact rock pieces.

② The value of the Hoek-Brown constant m_i for these intact rock pieces.

③ The value of the geological strength index GSI (for intact rock GSI=100).

1.4.4 Hoek-Brown failure and Mohr-Coulomb failure criteria

Many geotechnical software programs and design methods are written in terms of the Mohr-Coulomb failure criterion, therefore it may sometimes be necessary to determine equivalent angles of friction and cohesive strengths.

This is done by fitting an average linear relationship to the curve generated by solving the equation of the Hoek-Brown criterion fora range of minor principal stress values defined by $\sigma_t < \sigma_3 < \sigma_{3max}$.

The fitting process involves balancing the areas above and below the Mohr-Coulomb plot.

Relationships between major and minor principal stresses for Hoek-Brown and equivalent Mohr-Coulomb criteria.

Note that the value of σ_{3max} has to be determined for each individual case.

Chapter 2 Rock engineering design

Design approach:

(1) Deductive approach

As shown in Figure 2-1. Known conditions are used to deduce the behavior of a well defined problem.

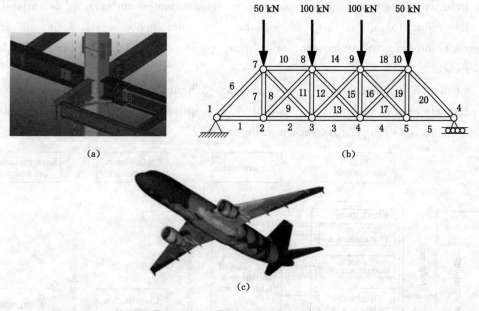

Figure 2-1 Engineering example

(2) Inductive approach

As shown in Figure 2-2. Limited observations, experience and engineering judgment are used to infer the behavior of a poorly-defined problem.

Figure 2-2 Engineering example

2.1 Introduction

2.1.1 Geotechnical design

Design evolves with process stages:
① Pre-feasibility;
② Feasibility;
③ Engineering design;
④ Construction;
⑤ Operation;
⑥ Closure & decommission.

Inductive design: managing geological/model/parameters uncertainty & variability.

By-product of uncertainty & variability is that geotechnical design is largely empirical; however many current and future projects includes conditions outside our experience.

Challenge is to use models to close the gap with our empirical knowledge.

Experts require shifting from qualitative/descriptive to quantitative mindset, "How to put numbers to Geology". As shown in Figure 2-3.

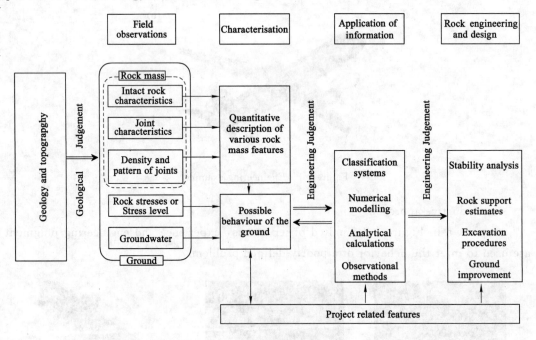

Figure 2-3　Geological, rock mechanics and rock engineering design flow chart

2.1.2 Rock engineering design

When is a rock engineering design acceptable?

There are no simple universal rules for acceptability nor are there standard factors of safety which can be used to guarantee that a rock structure will be safe and that it will perform adequately.

Each design is unique and the acceptability of the structure has to be considered in terms of the particular set of circumstances, rock types, design loads and end uses for which it is intended.

The responsibility of the geotechnical engineer is to find a safe and economical solution which is compatible with all the constraints which apply to the project.

Such a solution should be based upon engineering judgment guided by practical and theoretical studies such as stability or deformation analyses, when these analyses are applicable.

2.2 Data collection, data uncertainty and data variability

2.2.1 Introduction to data uncertainty and variability

Variability: an observable manifestation of heterogeneity of one or more physical parameters and/or processes.

Uncertainty: refers to the modeller's state of knowledge and strategy, and reflects the decision to recognise and address the observed variability in a qualitative and quantitative manner. As shown in Figure 2-4.

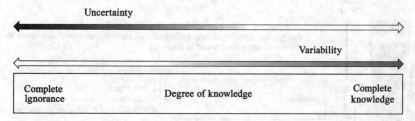

Figure 2-4　Data uncertainty and variability

In order to reduce engineering risk (e.g. reducing the potential for failure), it is necessary to consider both uncertainty (lack of knowledge) and variability (inherent randomness of a problem).

Uncertainty in rock engineering design:

There are various forms of uncertainty in rock engineering:

Geological uncertainty: unpredictability associated with the identification, characterization and interpretation of the site geology and hydrogeology.

Parameter uncertainty: absence of data for key parameters, spatial variability in rock/soil properties, and scale effects (e.g. intact rock vs. rock mass properties).

Model uncertainty: gaps in the scientific theory that is required to make predictions on the basis of causal inference.

Human uncertainty: subjectivity and measurement error, differing professional opinions. As shown in Figure 2-5.

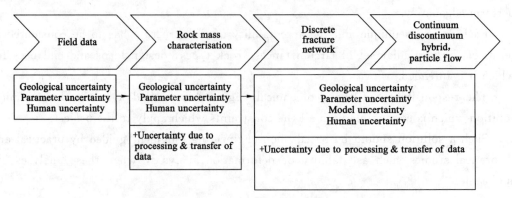

Figure 2-5 Uncertainty: from field data to numerical analysis

2.2.2 Variability and data collection methods (as shown in Figure 2-6)

A geotechnical study can be divided into three phases(as shown in Figure 2-7):

① Data collection;

② Data characterisation;

③ Data interpretation.

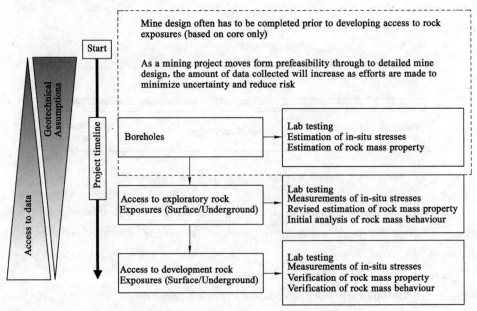

Figure 2-6 Rock engineering chart

(1) Collecting data

What happens when we are asked to map man-made rock exposures? Natural fractures and blast induced fractures. Can we differentiate the two?

Chapter 2　Rock engineering design

Figure 2-7　Collection steps of variable data

　　Road cut coincident with continuous, low friction (uplifted) bedding planes in shale.

　　Would scanline and/or window techniques be suitable mapping methods in this context?

　　Relationship between data collection and expected failure mechanism(s).

　　When working underground we have to think about the limitations which are imposed on data collection by design related features→size of the opening and exposed rock face. As shown in Figure 2-8.

Figure 2-8　Examples of underground works

　(2) Remote mapping techniques

　　LIDAR (light detection and ranging) works based on the principle of measuring the distance to a target by illuminating the target with light, often using pulses from a laser.

The remote mapping device is shown in Figure 2-9.

Figure 2-9　Remote mapping instrument

Field equipment: (a) Optech ILRIS-3D Laser scanner and (b) Canon EOS 30D digital camera with a $f=50$ mm lens. The rock map taken at the scene is shown in Figure 2-10.

(3) Classification of discontinuities

Principal structures-individual structures that affect all, or an important part, of the mine or a productive sector; treated deterministically.

Major structures-such as faults, dykes, shear zones and contacts, that may affect multiple underground excavations; treated deterministically.

Intermediate structures-joint sets, minor shears and bedding planes that impact localized areas (e.g. 2 or 3 production drifts); treated statistically.

Minor structures joints and veins that typically do not completely traverse a drift; treated statistically.

Discontinuities terminology:

Block forming joint: open joints and joints that are lightly cemented such that they form joints when the rock mass is deformed under low stress.

Defect: defects are cohesive or non-cohesive structures at the specimen scale that may influence the failure mode or the specimen strength. As shown in Figure 2-11.

(4) Influence of faults and shear zones

Off-setting the orebody.

Slip leads to a re-distribution of the field stresses and may be a source of seismicity.

Fretting or chimneying of friable material from within the fault or shear zone.

Isolation of blocks or wedges that may slide or fall.

Impact on fragmentation by large blocks or by friable material from within, or adjacent to, the fault or shear zone.

Inability to form satisfactory bolt and cable anchorages.

Chapter 2 Rock engineering design

Figure 2-10 Shooting rock samples at scene
(a) Murrin Lake; (b) Mt. Seymour; (c) Manning Park; (d) Saskatchewan crossing

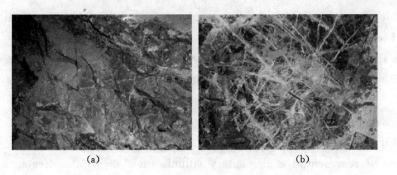

Figure 2-11 Cracks formed by blocks

Provide conduits for water.

2.2.3 Variability: discrete fracture networks

The discrete fracture network method provides a geologically realistic model based on the data we have measured in the field.

It is a stochastic (probabilistic) method, therefore we can generate multiple models, as shown in Figure 2-12.

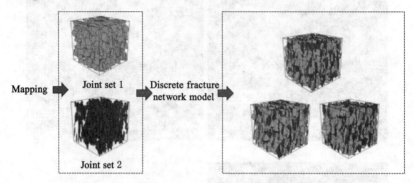

Figure 2-12　Discrete fracture model

Each one representing a geologically acceptable representation of reality. Each model looks different form the previous one, but all of them represent a geologically acceptable representation of reality.

2.2.4 Geotechnical data collection

Rock exposures (and rock cores) are utilised to obtain information on the engineering properties and structure of the rock mass.

In this lecture we will focus on collecting data on rock exposures (either surface or underground exposures).

Sampling problems to be considered:

What proportion of the rock mass should be surveyed to obtain satisfactory results? Concept of representative elementary volume (REV).

What degree of confidence can be placed on mean values of discontinuity properties determined using limited amounts of data?

There are no complete answers to these questions, although the use of statistical techniques, such as those developed by Priest and Hudson (1981) do provide valuable guidance.

What structure rating would you assign to this rock slope in terms of GSI? For example, Figure 2-13.

Concept of representative elementary volume (REV) → scale effects and rock mass properties.

Different types of structural features:

Bedding planes divide sedimentary rocks into beds or strata and represent interruptions in the course of deposition of the rock mass. Bedding planes are generally

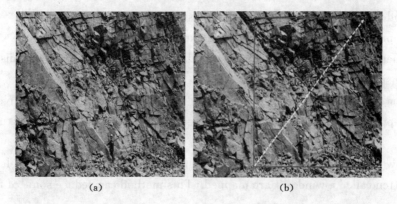

(a)　　　　　　　　　　　　(b)

Figure 2-13　Field rock pattern

highly persistent features.

Faults are fractures on which identifiable shear displacement has taken place. Faults may be pervasive features which traverse a mining area or they may be of relatively limited local extent on the scale of metres.

Shear zones are bands of material (up to several metres thick) in which local shear failure of the rock has previously taken place. They represent zones of stress relief in an otherwise unaltered rock mass throughout which they may occur irregularly.

Joints are the most common structural features in rocks. Joints are breaks of geological origin along which there has been no visible displacement. A group of parallel joints is called a joint set, and joint sets intersect to form a joint system. Joints may be open, filled or healed.

Veins, or cemented joints, are mineral infilling of joints or fissures. They have been found to have major influences on orebody capability and fragmentation (e. g. El Tenements mine). As shown in Figure 2-14.

Figure 2-14　Different types of structural features

(1) Collecting structural data (rock exposures)

The approaches used for mapping rock exposures include:

Spot mapping, in which the observer selectively samples only those discontinuities that are considered to be important.

Scanline (linear) mapping, in which all discontinuities intersecting a given sampling line are mapped. This is the basic technique used in mapping surface or underground exposures.

Window (cell or areal) mapping, in which all discontinuities within a selected area of the face, often called a window, are mapped. This method can reduce some of the biases in mapping.

① Scanline mapping

A scanline is a line set on the surface of the rock mass, and the survey consists of recording data for all discontinuities that intersect the scanline along its length.

The scanline preferentially intersects the longer discontinuity traces exposed at the rock face.

Due to the physical size limitations of the rock face it is likely that there will be somemaximum trace length that can be observed and measured.

From a practical point of view there may be a minimum trace length that can be observed and measured (trace length cut-off). As shown in Figure 2-15.

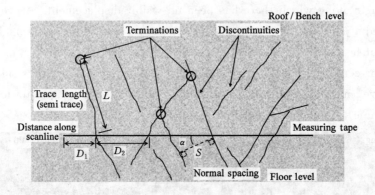

Figure 2-15 Schematic diagram of scanning line

Terzaghi correction: when orientation measurements are made, a bias is introduced in favour of those features which are perpendicular to the direction of surveying (DIPS manual, Rocscience, 2007).

② Window (areal) mapping

Window or areal mapping consists of recording data for all discontinuities that are contained within the defined cell. As shown in Figure 2-16.

③ Example of data included in a mapping sheet

Orientation: dip and dip direction are used to describe the attitude of the discontinuity

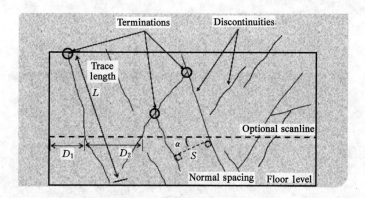

Figure 2-16 Window (areal) mapping

in space.

Type: It indicates the nature of the fracture being investigated (e.g. joint, bedding plane, foliation, and fault).

Aperture: perpendicular distance between adjacent walls of a discontinuity.

Infill: it describes the material that separates adjacent discontinuity walls. Presence of soft clay or other highly weathered materials and rock particles was considered.

Spacing: perpendicular distance between adjacent discontinuities. For a scanline survey it represents the progressive distance along the scanline, whilst in the case of window mapping it is usually given as the perpendicular distance between discontinuities inferred to belong to the same set.

Persistence: the trace length of a discontinuity as observed in the mapped exposure.

Termination: the nature of the termination of each discontinuity trace may also be measured (e.g. discontinuity trace terminates in rock, against another discontinuity, or it extends outside the mapped region).

Other parameters:

JRC is the roughness coefficient introduced by Barton (1973).

Roughness class represent the roughness profiles as described in ISRM (1981).

J_n, J_r, J_a are the parameters used in the NGI (Q-system) classification system (Barton et al., 1974).

(2) Case studies (Middleton mine: data capture and synthesis)

Discontinuity mapping at Middleton mine was undertaken by Dr. Davide Elmo and Dr. Zara Flynn (Research Assistant at Camborne School of Mines).

Mapping four faces of a pillar located on Level 1, including a detailed window mapping of 15 m×2 m panels located at the base of the pillar faces. As shown in Figure 2-17.

Orientation bias: the probability of a fracture appearing in a window depends on the relative orientations of the fracture and the window. Fractures striking perpendicular to the window will be truly sampled, whilst all others will be under-represented. To account

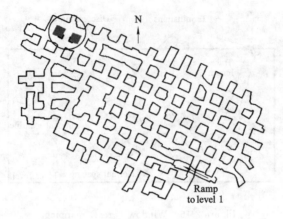

Figure 2-17 Mapping four
faces of a pillar located on level 1

for this, at least two perpendicular windows were mapped. In the case of the rectangular windows mapped at Middleton (14 m long × 2 m high), it was expected that fractures striking parallel to long axis could be under-represented (e. g. if their average spacing was greater than the window height), hence the superimposition of circular windows within the mapped rectangular panels.

Size (length) bias: longer fractures are more likely to intersect a sampling plane. This is an important consideration when deriving radius distribution from trace length distribution.

Truncation bias: trace lengths below a specific cut-off value are not recorded. Trace lengths below 0.5 m long were not recorded when mapping, therefore, when fitting distributions, trace lengths less than 0.5 m were discarded accordingly.

Censoring bias: if the end points of a trace cannot be seen, only a lower bound estimate of its size can be made. The minimum size of the sampling window should be carefully chosen to reflect the fracture trace lengths. As shown in Figure 2-18.

Histograms of trace lengths measured for sets 1a, 1b, 2 and 3.

The fractures from sets 2a and 2b were initially analysed separately, and appeared to conform to the same radius distribution. The same approach was followed for sets 3a and 3b.

While sets 2 and 3 (combined) showed an essentially lognormal trace length distribution, as often encountered in the field, sets 1a and 1b showed a bimodal distribution due to the relatively high number of long fractures. As shown in Figure 2-19.

Defining fracture size distributions:

The derivation of fracture size distribution is critical to any modelling, however size is often among the most difficult parameters to constrain.

Any measurements relating to fracture size are actually measurements of the trace a

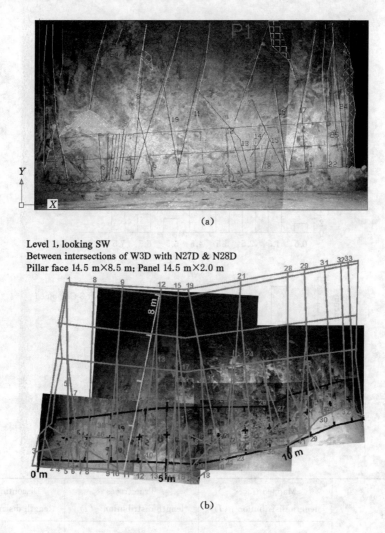

Figure 2-18 Example of rock scene

fracture or fault make with a geological surface or mining exposure (chord to a "disc").

Need to determine the underlying fracture size distribution that results in the observed trace length distribution. As shown in Figure 2-20 and Table 2-1.

There are a number of ways that can be done:

Analytical method;

Scaling laws;

Manual simulated sampling;

Automated simulated sampling.

Model need to be validated against field observations remember relationship between scale at which data were collected and model resolution/scale. 2D modelling of rock mass strength based on the models generated for Middleton mine[31].

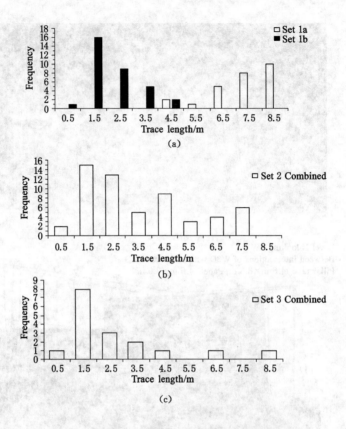

Figure 2-19　Index change polygon chart

Table 2-1　　　　　　　　　Model size and parameters

Select	Mapped traces length distribution $g(I)$		True traces length distribution $g(I)$		Discontinuity traces length distribution $g(D)$	
	Log-Normal		Log-Normal		Log-Normal	
Sets	u_g	σ_g	u_f	σ_f	u_D	σ_D
1a	38.9	9.0	32.2	12.0	38.9	9
1b	3.3	0.6	2.6	0.9	3.3	1.6
2a	3.7	1.2	3.2	1.4	3.2	1.4
2b	3.7	1.2	3.2	1.4	3.2	1.4
3a	3.7	1.5	3.4	1.7	3.7	1.5
3b	3.7	1.5	3.4	1.7	3.7	1.5

Chapter 2 Rock engineering design

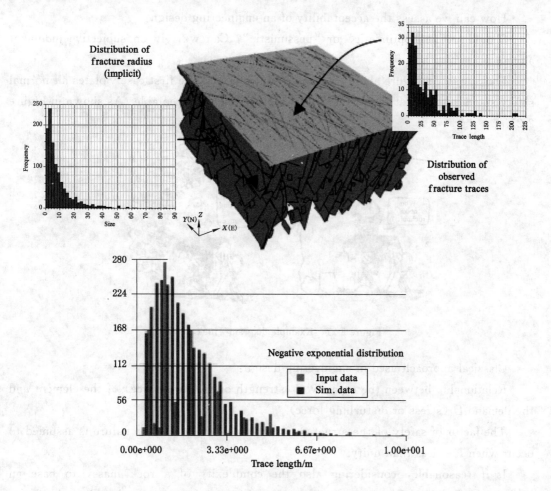

Figure 2-20 Numerical model established

2.3 Factor of safety and probability of failure

(1) Many of the details of rock mass behavior are unknown and unknowable→there will always be a degree of uncertainty in any numerical model.

Analysis of rock mass response involves different scales. Is it really necessary and desirable to include all features and details of rock mass response mechanisms into one model?

Modelling requires the real problem be idealized and simplified:

In that empire, the art of cartography attained such perfection that the map of a single province occupied the entirety of a city, and the map of the empire, the entirety of a Province. In time, those unconscionable maps no longer satisfied, and the cartographers guilds struck a map of the empire whose size was that of the Empire, and which coincided point for point with it. On exactitude in science, J. L. Borges[32].

How can we assess the acceptability of an engineering design?

Should we be "optimistic" or "pessimistic"? Can we rely on subjective judgment alone?

Looking at the figure above, we could conclude that the first case violates all normal safety standards, whilst the second one is economically unacceptable. As shown in Figure 2-21.

Figure 2-21 Example diagram(Hoek, 2007)

Classical approach used in engineering design:

Relationship between the capacity C (strength or resisting force) of the element and the demand D (stress or disturbing force).

The factor of safety of the structure is defined as $F=C/D$ and failure is assumed to occur when F is less than unity.

Is it reasonable, considering also the complexity of a rock mass, to base an engineering design decision on a single calculated factor of safety, or should we decide for an approach giving a more rational assessment of the risk(s)?

(2) Covariance of modelling parameters:

Modelling parameters generally are not independent variables.

For example: cohesive strength generally decreases as the friction angle increases, and vice versa.

For rock engineering design a probability analysis should be such that sampling of the PDF curves accounts for covariance of the modelling parameters (data variability→model uncertainty). As shown in Figure 2-22.

Sensitivity analysis (as shown in Figure 2-23):

Series of calculations in which each significant parameter is varied systematically over its maximum credible range in order to determine its influence upon the factor of safety.

This approach can provide a useful means of exploring a range of possibilities and reaching practical decisions on some difficult problems[33,34].

(3) Example (sensitivity analysis):

Consider the case of a pattern of rock bolts which are designed to hold up a slab of

Chapter 2 Rock engineering design

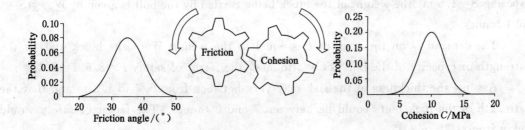

Figure 2-22 Covariance of modelling parameters

Figure 2-23 Sensitivity analysis

rock in the back of an excavation. As shown in Figure 2-24.

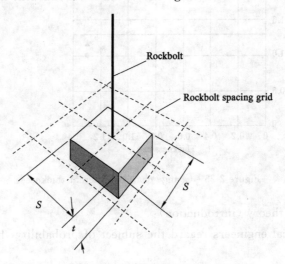

Figure 2-24 Rock bolt pattern

Slab of thickness t being supported by one rock bolt in a pattern spaced on a grid spacing of $S \times S$.

Unit weight the broken rock $g = 2.7$ tonnes/m^3, thickness of the slab $t = 1$ m, grid

· 35 ·

spacing $S=1.5$ m, the weight of the block being carried by the bolt is given by $W=gtS^2=$ 6.1 tonnes.

The demand D on the rock bolt is equal to the weight W of the block and, if the strength or capacity of the bolt is $C=8$ tonnes, the factor of safety $F=8/6.1=1.3$.

Assume the thickness of the slab may vary between from 0.7 to 1.3 m and that the strength of the rock bolts could lie between 7 and 9 tonnes. The factor of safety would vary from 0.88 to 2.12.

Maximum thickness & minimum bolt capacity→minimum FoS (0.88).

Decreasing resisting forces & increasing disturbing forces.

Minimum slab thickness & maximum bolt capacity→maximum FoS (2.12).

Increasing resisting forces & decreasing disturbing forces.

By decreasing the bolt spacing from 1.5 to 1.4 m we would have (the changes are shown in Table 2-2 and Figure 2-25).

Table 2-2 Reduced spacing variation table

Bolt spacing/m	Thickness/m	Bolt capacity/tonnes	FoS
1.4	0.7	9	2.43
1.5	1.3	7	0.88

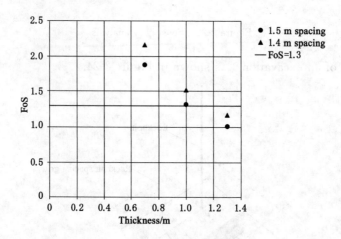

Figure 2-25 Change chart of different thickness

(4) Probability theory (introduction):

Most geotechnical engineers regard the subject of probability theory with doubt and suspicion.

At least part of the reason for this mistrust is associated with the language which has been adopted by those who specialise in the field of probability theory and risk assessment.

Important definitions:

Random variables: it can represent parameters such as the angle of friction of rock joints, the uniaxial compressive strength of rock specimens, the inclination and orientation of discontinuities in a rock mass and the measured in situ stresses in the rock surrounding an opening do not have a single fixed value but may assume any number of values[35].

We know that it is very difficult (if not impossible) to predict exactly what the value of one of these parameters will be at any given location. Hence, these parameters are described as random variables.

Probability distribution: a probability density function (PDF) describes the relative likelihood that a random variable will assume a particular value. Example: the random variable is continuously distributed (i. e., it can take on all possible values). The area under the PDF is always unity (=1 or 100%).

An alternative way of presenting the same information is in the form of a cumulative distribution function (CDF), which gives the probability that the variable will have a value less than or equal to the selected value.

The CDF is the integral of the corresponding probability density function, i. e., the ordinate at x_1 on the cumulative distribution is the area under the probability density function to the left of x_1.

Note the $f_x(x)$ is used for the ordinate of a PDF while $F_x(x)$ is used for a CDF. As shown in Figure 2-26 and Figure 2-27.

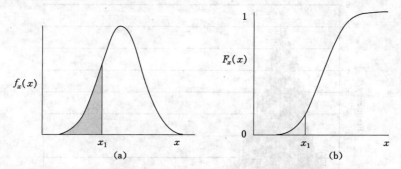

Figure 2-26 Probability density function and its integral(Hoek,2007)
(a) Probability density function(PDF);(b) Cumulative distribution function(CDF)

It is desirable to include as many samples as possible in any set of observations but, in geotechnical engineering, there are serious practical and financial limitations to the amount of data which can be collected.

Therefore, it is often necessary to make estimates on the basis of judgement, experience or from comparisons with results published by others.

There is not a good reason for not using probability theory in geotech engineering:

Difficulties associated with limited data... But useful results can still be obtained from very limited data.

Need to know and understand the mathematics involved in all of these probability distributions…But software programs are now commercially available (but be aware of the

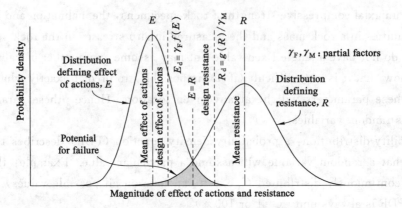

Figure 2-27 Schematic of the principles of limit state design (Becker, 1996)

black-box syndrome).

One of the most common graphical representations of a probability distribution is a histogram in which the fraction of all observations falling within a specified interval is plotted as a bar above that interval.

The sample mean (or expected value or first moment) indicates the centre of gravity of a probability distribution. As shown in Figure 2-28.

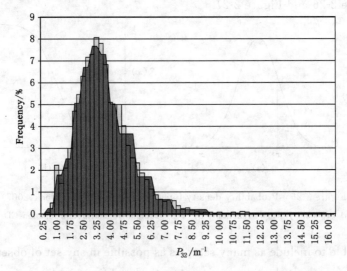

Figure 2-28 Histogram representation of probability distribution

Example: a typical application would be the analysis of a set of results $x_1, x_2, \cdots, x_n$ from uniaxial strength tests carried out in the laboratory[36-38]. Assuming that there are n individual test values x_i, the mean is given by:

$$\bar{x} = \frac{1}{n} \sum_{i=1}^{n} x_i$$

The sample variance s^2 or the second moment about the mean of a distribution is defined as the mean of the square of the difference between the value of x_i and the mean value.

$$s^2 = \frac{1}{n-1}\sum_{i=1}^{n}(x_i - \bar{x})^2$$

The standard deviations is given by the positive square root of the variance s^2. For a normal distribution, about 68% of the test values will fall within an interval defined by the (mean ± one standard deviation) while approximately 95% of all the test results will fall within the range defined by the (mean ± two standard deviations).

Smaller standard deviations indicate a tightly clustered data set while larger standard deviation indicate a data set in which there is a large scatter about the mean.

The coefficient of variation (COV) is the ratio of the standard deviation to the mean, i.e. COV=σ/μ. COV is dimensionless and it is a particularly useful measure of variability (example: COV=0.05→small variability while COV=0.25→large variability).

Normal distribution:

The normal (or Gaussian) distribution is the most common type of probability distribution function and the distributions of many random variables conform to this distribution.

It is generally used for probabilistic studies in geotechnical engineering unless there are good reasons for selecting a different distribution.

We need to define the values governing the distribution, which are the true mean (μ) and true standard deviation (σ). Generally, the best estimates for these values are given by the sample mean and standard deviation, determined from a number of tests or observations. Therefore $\sigma = s$.

Other distributions:

Beta distributions(Harr, 1987) are very versatile distributions which can be used to replace almost any of the common distributions and which do not suffer from the extreme value problems discussed above because the domain (range) is bounded by specified values.

Exponential distributions are sometimes used to define events such as the occurrence of earthquakes or rockbursts or quantities such as the length of joints in a rock mass.

Lognormal distributions are useful when considering processes such as the crushing of aggregates in which the final particle size results from a number of collisions of particles of many sizes moving in different directions with different velocities. Such multiplicative mechanisms tend to result in variables which are lognormally distributed as opposed to the normally distributed variables resulting from additive mechanisms.

Weibul distributions are used to represent the lifetime of devices in reliability studies or the outcome of tests such as point load tests on rock core in which a few very high values may occur.

Consider a problem in which the factor of safety depends upon a number of random variables such as the cohesive strength, the angle of friction and the acceleration a due to earthquakes or large blasts.

Assume that the values of these variables are distributed about their means in a manner which can be described by one of the continuous distribution functions such as the normal distribution described earlier, the problem is how to use this information to determine the distribution of factor of safety values and the probability of failure.

The Monte Carlo method:

It uses random or pseudo-random numbers to sample from probability distributions and, if sufficiently large numbers of samples are generated and used in a calculation (e. g. factor of safety), a distribution of values for the end product will be generated.

Note that the Monte Carlo requires that the distribution of all the input variables should either be known or that they be assumed. When no information on the distribution is available it is usual to assume a normal or a truncated normal distribution.

The Latin Hypercube:

This is a sampling technique that gives comparable results to the Monte Carlo technique, but with fewer samples.

The method is based upon stratified sampling with random selection within each stratum. Typically, an analysis using 1 000 samples obtained by the Latin Hypercube technique will produce comparable results to an analysis using 5 000 samples obtained using the Monte Carlo method.

Note that both the Monte Carlo and the Latin Hypercube techniques require that the distribution of all the input variables should either be known or that they be assumed.

When no information on the distribution is available, it is usual to assume a normal or a truncated normal distribution. As shown in Figure 2-29.

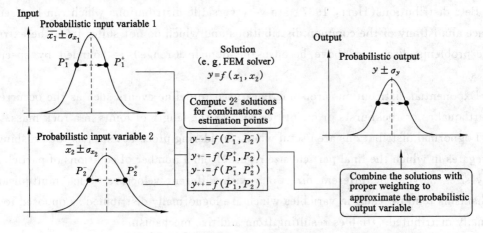

Figure 2-29 Output and input images

The generalised point estimate method:

It can be used for rapid calculation of the mean and standard deviation of a quantity such as a factor of safety which depends upon random behaviour of input variables.

To calculate a quantity such as a factor of safety, two point estimates are made at one standard deviation on either side of the mean($\mu \pm \sigma$) from each distribution representing a random variable[39,40].

The factor of safety is calculated for every possible combination of point estimates, producing 2^n solutions where n is the number of random variables involved. The mean and the standard deviation of the factor of safety are then calculated from these 2^n solutions.

This technique does not provide a full distribution of the output variable, but it is very simple to use for problems with relatively few random variables and is useful when general trends are being investigated.

Chapter 3　Rock excavation methods

In mining and geotechnical engineering, both must face the problem of rock excavation. In order to excavate rock mass more efficiently, people have designed a variety of methods to solve the difficult problem of excavating complex rock mass. The application of these technologies in practice has greatly strengthened the safety of construction, the speed of excavation and the maximization of benefits. It has brought great benefits to the society. This chapter will give a detailed introduction to the common methods of geotechnical excavation.

3.1　Introduction

3.1.1　Excavations in rock

The fundamental objective of the excavation process is to remove rock material.

① Material is removed to create an opening (tunnel, stope, et al.).

② Material is excavated to extract ore.

In general terms, anytime rock material is removed, it is necessary to introduce additional fracturing and fragmentation of the rock material.

Key aspects of the excavation process include:

① Overcoming the peak strength of the intact rock.

② The in-situ block size defined by the natural occurring joints must be changed to meet the planned fragmentation size (very important in block caving).

③ Conditions 1 and 2 above are met by introducing energy into the system (rock mass) by means of different excavation techniques.

The tensile strength of the rock is generally $1:10^{th}$ of the compressive strength.

Breaking the rock in tension requires $1:100^{th}$ of the energy required to break the rock in compression.

Need to optimise the use of energy to remove a unit volume of rock (energy is expressed as $J\ m^{-3}$).

There are two ways of introducing energy into the rock mass:

① Blasting, by means of which energy is introduced in the rock mass in large quantities over a very short period of time.

② Machines, by means of which energy is introduced in the rock mass in smaller quantities but continuously.

3.1.2 Mechanics of rock breakage

The rock breaking process can be classified into three major groups[41,42]:

Primary, application of a force by means of a hard indenter to a free rock face much larger than the indenter.

This generates chips which are of a size similar to that of the indenter at the sides of the indenter and a pulverised zone immediately below the indenter.

Secondary, application of forces inside a hole near to the rock face.

The forces inside the hole generate tension at the sides of the hole which produces cracks running to the free surface.

Tertiary, application of forces from more than one side to a free surface.

(1) Primary breakage processes

Impact or hammering: dynamic forces are applied.

Percussive drilling: application of a hard indenter to the bottom of a hole. The force is applied from one side only and the bottom of the hole is the free face.

Button type cutters for raise and tunnel borers.

The buttons are loaded slowly (quasi-statically) and are moved away to be re-applied elsewhere.

That is, indexing occurs by rolling to the next button.

Repeated applications over a large surface area maintain the flat face.

Disc type cutters for raise and tunnel borers: hard indenter indexed by rolling. Forces at a point in the rock rise very slowly[43].

Drag-bit: a hard indenter forced onto the rock and indexed by dragging across the surface.

Diamond bits, a very hard surface and very small indenter dragged across the surface.

The real breaking is done by the force thrusting the diamonds against the rock. Diamonds produce very small fragments because they are small indenters.

(2) Secondary breakage processes

Wedging, by means of wedges driven into a hole which produces crack.

Blasting, use of explosive generates a pulverised zone through compression but the real breaking process is by driving tensile cracks.

(3) Tertiary breakage processes(as shown in Figure 3-1, Figure 3-2 and Figure 3-3)

The tertiary breakage processes are closely related to breaking the rock in tension.

Think about how loading a sphere by diametrically opposed forces causes a uniform tensile stress across the diametrical plane.

Breaking boulders by impact or mud blasting.

Crushing.

Milling.

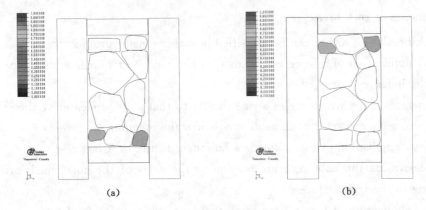

Figure 3-1 Example of modelling of tertiary processes

Figure 3-2 Excavations in rock

Figure 3-3 Rock excavation equipment

3.2 Drill & blast

Drilling and blasting is the most common method of rock breaking.

Application of explosive in the rock is carried out by means of drilling holes (also knows as shot holes or blast holes) depending upon their length and diameter.

Blast design is an iterative process, for which important factors need to be considered:

① Required fragmentation.

② Production.

③ Muck pile shape (mostly for surface mining).

These factors in turn control:

① Optimal drill-hole diameter.

② Depth and inclination.

③ Sub-drilling.

④ Explosive types.

⑤ Detonation timing.

3.2.1 Drill & blast: mechanism of rock breakage(as shown in Figure 3-4)

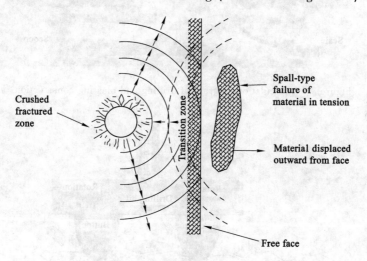

Figure 3-4 Mechanism of rock breaking by drilling and blasting

The detonation of a charge in a blast hole induces compressive strain waves into the surround rock.

As a result, a radial crack pattern around the blast hole is generated. These cracks typically extend at a distance of 4 blast hole diameters into the rock.

A reflected tensile stress wave is generated when the compressive forces encounters a free face, which tends to open the cracks which are parallel to sub parallel to the free face.

High pressure gas preferentially enters these open cracks and generates a force acting outwards, towards the free face. As shown in Figure 3-5.

3.2.2 Rock drill ability: crushing mechanism

(1) Around the contact of the button a new state of stress is induced in the rock, where four important destruction mechanisms can be distinguished (Figure 3-6 shows the process of rock breaking)[44-46]:

① Under the bit button a crushed zone of fine rock powder is formed (impact).

② Starting from the crushed powder zone, radial cracks are developed (induced tensile stress).

③ When stress in the rock is high enough (if enough cracks exist ± parallel to the

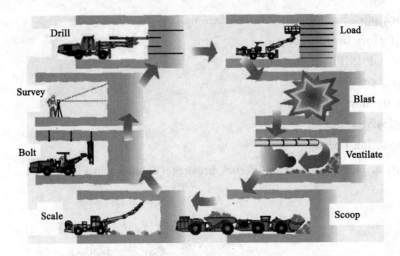

Figure 3-5 Circulation diagram of drilling and blasting construction

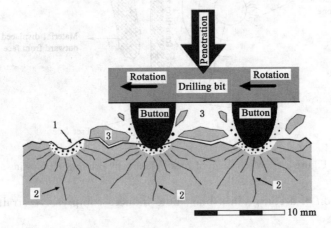

Figure 3-6 Rock fragmentation process

bottom of the borehole), larger fragments of the rock can be sheared off between the button grooves (shear stress).

④ addition to the mechanisms above stress is induced periodically (dynamic process).

(2) Rock drill ability:

Drill ability is a term used in construction to describe the influence of a number of parameters on the drilling rate (drilling velocity) and the tool wear of the drilling rig.

Apart from technical parameters, especially the geological parameters will basically influence the drilling performance and the wear of the drilling rig.

The drill ability of a given rock is determined by a number of factors. Amongst these are the mineral composition, the grain size and the brittleness. Sometimes drill ability is also referred back to the rock compressive strength or hardness.

Tool wear is often proportional to drill ability, though it also depends on how abrasive the rock is. The structural composition of the drill is shown in Figure 3-7.

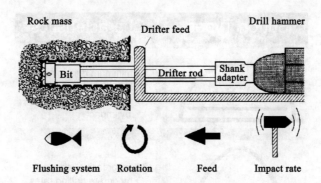

Figure 3-7 Structural composition of drill bits(Thuro,1997)

(3) Rock drill ability: the drilling bit.

The drilling bit is the part of the rig which carries out the crushing work.

Typical button drill bits with six, seven, eight and nine buttons and different flushing systems mainly used in hard rock.

The bit consists of a carrier holding the actual drilling tools: buttons of hard metal (e. g. wolfram carbide with a cobalt binder, MOHS hardness $9\frac{1}{2}$). As shown in Figure 3-8, Figure 3-9, Figure 3-10 and Figure 3-11.

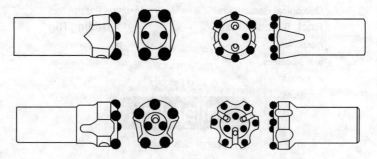

Figure 3-8 Shape of drill bit

(4) Rock drill ability: importance of geological parameters.

Geological influences can have a significant impact on drilling velocity and bit life.

There are several geological influences, including:

Anisotropy orientation of discontinuities related to the direction of testing or drilling.
Spacing of discontinuities.
Mineral composition (equivalent quartz content).
Pore volume(porosity of the micro fabric).

Read more here: Drillability prediction-geological influences in hard rock drill and blast tunnelling

3.2.3 Rock drill ability: anisotropic rock mass

Rock properties and drilling rates are highly dependent on the orientation of weakness

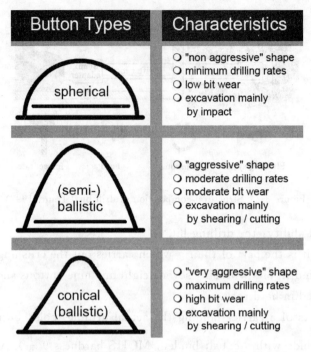

Figure 3-9 Types and characteristics of buttons

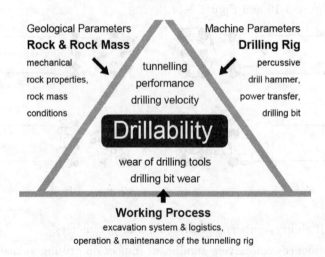

Figure 3-10 Classification of rock drill ability

planes (e. g. foliation) related to the direction of testing or drilling.

Drilling at right angles to the orientation of foliation:

Rock material is compressed at right angles but sheared parallel to it.

Cracks will develop radial to compression, the cracks parallel to the bottom of the borehole will be used for chipping[47,48].

Usually in this case the highest drilling velocities are obtained, because of the favorable orientation of the foliation planes.

Drilling is controlled by the shear strength of the foliated rock material.

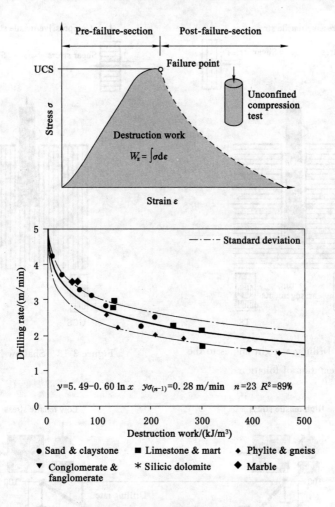

Figure 3-11 Wear curves of drill bits and standard deviation

The minimum destruction work causes large sized chips and a maximum drilling performance. As shown in Figure 3-12.

Drilling parallel to the orientation of foliation:

Compression also is parallel but shear stress is at right angles.

Fewer cracks will develop for reasons of higher strength at right angles to foliation.

Drilling is controlled by the tensile strength parallel to the foliation producing small-sized fragments and a minimum drilling performance[49,50].

It is certain, that in the parallel case, rock properties are the highest and drilling rates are low. As shown in Figure 3-13.

As a further result of anisotropy, problems may also occur when drilling direction is diagonal to the tunnel axis.

When the angle between drilling and tunnel axis is acute-angled, drifter rods are deviated into the dip direction of foliation, if obtuse-angled, into the normal direction of foliation. As shown in Figure 3-14.

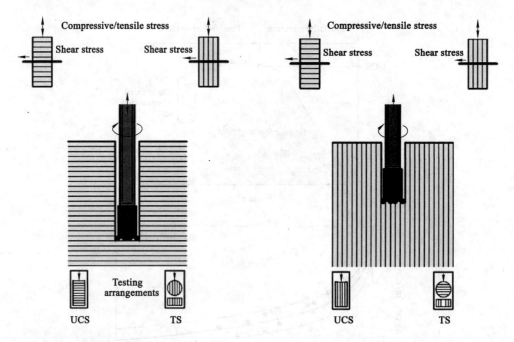

Figure 3-12　Drilling at right angles to the orientation of foliation

Figure 3-13　Shallow hole drilling

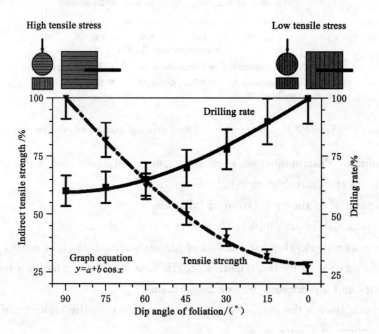

Figure 3-14　Drilling rate and tensile strength plotted against the orientation of foliation

Drill tracks may be seen as curves and produce distinct borehole deviation and a "geologically caused" over break.

Drilling rate and tensile strength plotted against the orientation of foliation. As shown in Figure 3-15.

Chapter 3　Rock excavation methods

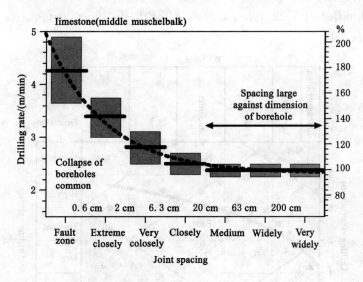

Figure 3-15　Relationship between drilling rate and discontinuous spacing

3.2.4　Rock drill ability: spacing of discontinuities

(1) Drilling rates are also dependent on spacing of discontinuities in rock mass.

Very closed spaced discontinuities→borehole instability, causing hole collapses and time consuming scaling of the established blast hole (fault zones are a an extreme example). As shown in Figure 3-16.

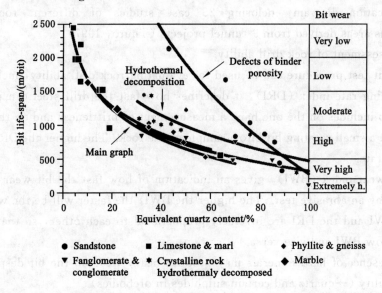

Figure 3-16　Relationship between drilling speed and equivalent quartz content

(2) Rock drill ability: mineral composition.

Tool wear is predominantly a result of the mineral content harder than steel (Mohs hardness ca. 5.5), especially quartz[51,52] (Mohs hardness of 7).

Bit life of different rock types correlated with its equivalent quartz contents. As shown in Figure 3-17.

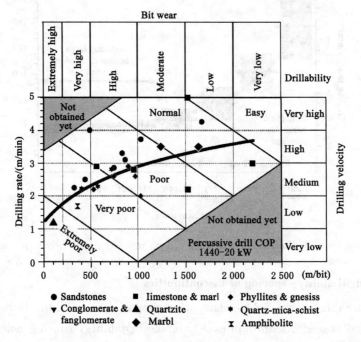

Figure 3-17 Classification of rock drill ability

Classification diagram enclosing 25 case studies of different rock types or homogeneous areas derived from 9 tunnel projects (Thuro, 1997).

(3) Assessment of rock drill ability.

Different test procedure can be used for evaluating rock drill ability, including:

① Drilling rate index (DRI), it describes how fast the drill steel can penetrate the rock. It also includes on the one hand a measurement of brittleness and on the other hand drilling with a small rotating bit into a sample of the rock. The higher the DRI, the higher penetration rate.

② Bit wear index (BWI), gives an indication of how fast the bit wears down. It is determined by an abrasive test. The higher the BWI, the faster will be the wear. In most cases the BWI and the DRI are inversely proportional to each other, so that a high DRI will give a low BWI and vice versa.

The presence of hard minerals may produce heavy wear on the bit despite relatively good drillability (→quartz and certain sulphides in orebodies).

3.2.5 Drill & blast: drilling patterns

Various drilling patterns have been developed for blasting solid rock faces, such as:

① Burn cut.

② Wedge cut or V cut.

③ Pyramid or diamond cut.

④ Drag cut.

⑤ Fan cut.

In today's drill and blast tunnelling in which multi-boom drilling machines are used, the most convenient method for creating the initial void is the burn cut.

(1) Drill &·blast: burn cut

The burn cut consists of a series of parallel holes drilled closely spaced at right angles to the face. One hole or more at the centre of the face are uncharged.

The uncharged holes are often of larger diameter than the charged holes and form zones of weakness that assist the adjacent charged holes in breaking out the ground.

Since all holes are at right angles to the face, hole placement and alignment are easier than in other types of cuts.

The burn cut is particularly suitable for use in massive rock such as granite, basalt etc. As shown in Figure 3-18.

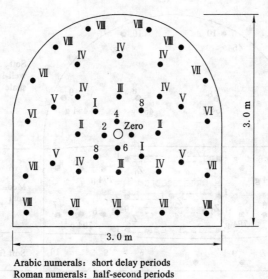

Arabic numerals: short delay periods
Roman numerals: half-second periods

Figure 3-18 Schematic diagram of burning cut hole

(2) Drill &·blast: wedge cut(as shown in Figure 3-19)

Blast hole are drilled at an angle to the face in a uniform wedge formation so that the axis of symmetry is at the centre line of the face.

This pattern is such that it displaces a wedge of rock out of the face in the initial blast.

The wedge is subsequently enlarged to the full width of the drift in subsequent blasts (each blast being fired with detonators of suitable delay time)[53,54].

① The wedge cut is particularly suited to large size drifts, which have well laminated or fissured rocks.

② Hole placement should be carefully preplanned and the alignment of each hole

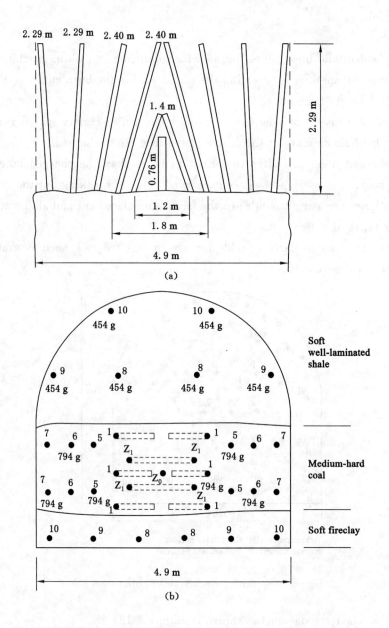

Figure 3-19 Sketch map of wedge cut hole

should be accurately drilled.

(3) Drill & blast: pyramid or diamond cut (as shown in Figure 3-20)

Variation of the wedge cut where the blast holes for the initial cavity may have a line of symmetry along horizontal axis as well as the vertical axis.

(4) Drill & blast: drag cut (as shown in Figure 3-21)

The drag cut is particularly suitable in small sectional drifts where a pull of up to 1 m is very useful.

(5) Drill & blast: fan cut (as shown in Figure 3-22)

Chapter 3 Rock excavation methods

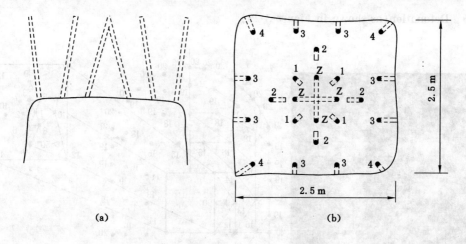

Figure 3-20 Pyramid or diamond cut

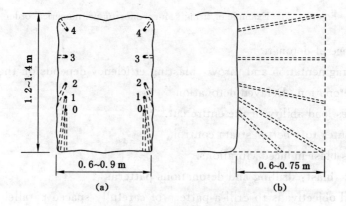

Figure 3-21 Drill & blast: drag cut

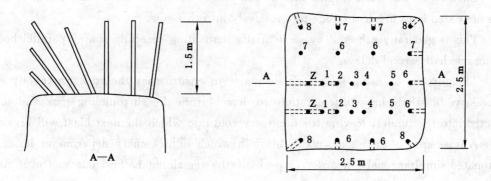

Figure 3-22 Drill & blast: fan cut

The fan cut is one-half of a wedge cut and is applicable mainly where only one machine is employed in a narrow drive.

Generally the depth of pull obtainable is limited to 1.5 m.

3.2.6 Detonation(as shown in Figure 3-23)

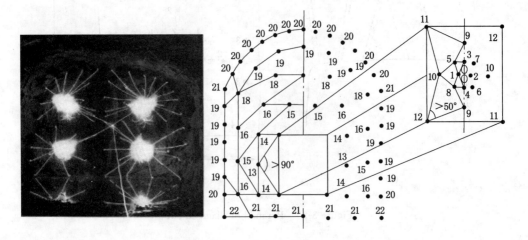

Figure 3-23　Example of firing sequence for tunnel numerical order

(1) Sequence of detonation

For both fragmentation and throw, blasting efficiency depends on the delay sequence of blast hole detonation. Delayed detonation[55-57]:

① Improves load ability of the entire cut.

② Contributes to a better strata control.

③ Reduces blast-induced vibrations.

(2) Drill & blast: drilling and detonations patterns

The overall objective is to drill a pattern of carefully spaced parallel holes which are then charged with powerful explosive and detonated sequentially using millisecond delays.

Once a void has been created for the full length of the intended blast depth or pull, the next step is to break the rock progressively into this void.

This is generally achieved by sequentially detonating carefully spaced parallel holes, using one-half second delays.

The purpose of using such long delays is to ensure that the rock broken by each successive blasthole has sufficient time to detach from the surrounding rock and to be ejected into the tunnel, leaving the necessary void into which the next blast will break.

A final step is to use a smooth blast in which lightly charged perimeter holes are detonated simultaneously in order to peel off the remaining half to one metre of rock, leaving a clean excavation surface[58-60].

Development of a burn cut using millisecond delays(as shown in Figure 3-24).

Use of half-second delays in the main blast and smooth blasting of the perimeter of a tunnel (with burn-cut). As shown in Figure 3-25.

(3) Example of good blasting(as shown in Figure 3-26)

Results achieved using well designed and carefully controlled blasting in a 6 m

Chapter 3 Rock excavation methods

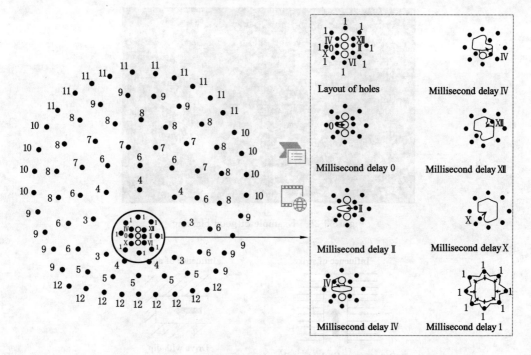

Figure 3-24 Example of sequence of detonation with burn cut

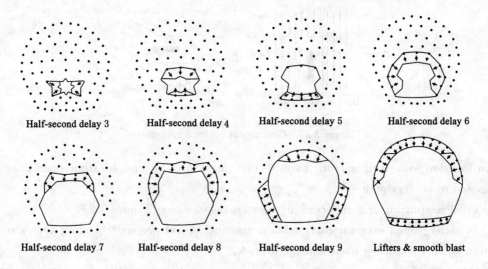

Figure 3-25 Example of sequence of detonation

diameter tunnel in gneiss in the victoria hydroelectric project in Sri Lanka.

Note that no support is required in this tunnel as a result of the minimal damage inflicted on the rock.

3.2.7 Classification of blasting modes

(1) Blasting: anisotropic rock mass(as shown in Figure 3-27)

Blasting conditions are often related to drilling. If the tunnel axis is parallel to the

· 57 ·

Figure 3-26 Example of good blasting

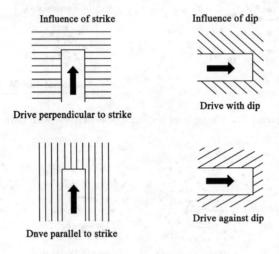

Figure 3-27 Blasting: anisotropic rock mass

main foliation set, it is safe to assume that main drilling and blasting conditions are supposed to be very poor[61,62].

(2) Pre-splitting and underground blasting(as shown in Figure 3-28)

In underground excavations is seldom practical to use pre-split blasting (see also the section about controlling blast damage in rock).

Pre-split blasting may be used, however, in the case of a benching operation underground.

In a pre-split blast, the closely spaced parallel holes (numbered 9, 10 and 11) are detonated before the main blast instead of after, as in the case of a smooth blast.

Since a pre-split blast carried out under these circumstances has to take place in almost completely undisturbed rock which may also be subjected to relatively high induced stresses, the chances of creating a clean break line are not very good.

The cracks, which should run cleanly from one hole to the next, will frequently veer off in the direction of some pre-existing weakness such as foliation.

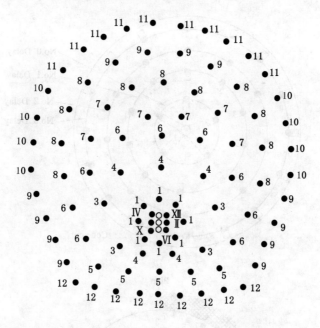

Figure 3-28　Perforation arrangement for pre splitting blasting

(3) Drill & blast: shaft sinking(as shown in Figure 3-29)

In mining, shafts are used for transportation of ore, refill, personnel, equipment, air, electricity, ventilation et al.

An important requirement in shaft sinking is to provide optimum fragmentation of the rock so that it can be cleared quickly from the congested shaft-face area.

Because blasting operation is carried out against gravity, the scatter of the broken rock is confined in the shaft.

It is common to use generous distribution of explosives throughout the rock using a large number of small diameter (35~42 mm) shot holes.

The number of holes N required for sinking a shaft of cross sectional area A in m^2 is given by: $N=2.5A+22$.

Drilling patterns include the cone cut and bench cut.

3.2.8　Blasting in underground mines

(1) Blasting in underground mines: long-hole ring blasting

The excavation work for underground mines is usually divided into two broad categories:

① Development, which involves tunneling, shaft sinking, cross cutting, raising, etc. so that the ore material are easily accessible and transportable after excavation.

② Production, which can be further subdivided into two categories, short-hole and long-hole blasting.

There are three long hole blasting systems: Ring blasting; Bench blasting; Vertical crater retreat (VCR).

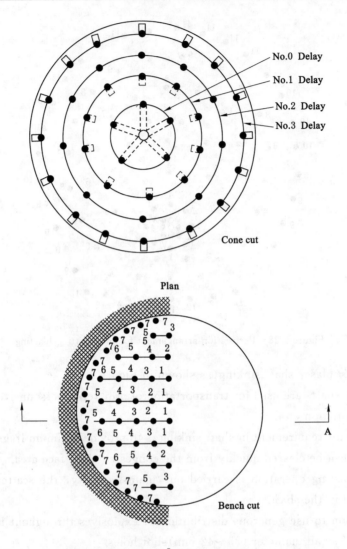

Figure 3-29 Drill & blast: shaft sinking

① Blasting in underground mines: long-hole ring blasting
As shown in Figure 3-30.
② Blasting in underground mines: long-hole bench blasting
As shown in Figure 3-31.
Bench blasting is essentially equivalent to surface excavation.
A development heading is first excavated at the top sublevel to provide drilling space. Then depending on thickness of orebody and/or availability of drilling machinery, either Vertical or horizontal blast holes are drilled to increase the height of the excavation.
③ Blasting in underground mines: long-hole VCR blasting
As shown in Figure 3-32.
Vertical or subvertical blast holes are drilled downward from the top level to the bottom level.

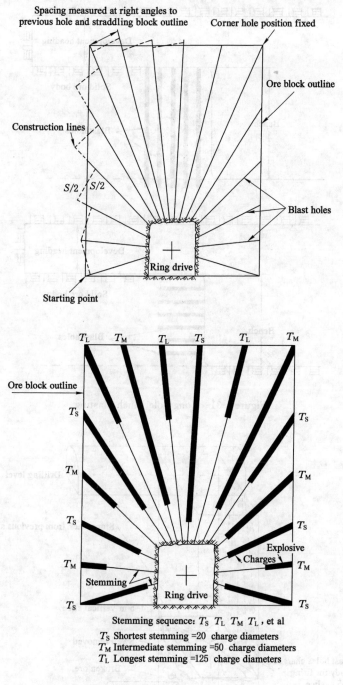

Figure 3-30 Long-hole ring blasting

A cuboid of orebody can be excavated from the lower level upward by a number of horizontal slices using the same blast holes.

Spherical charges should be placed to obtain the maximum cratering effect.

(2) Blasting in underground mines: short-hole blasting (as shown in Figure 3-33)

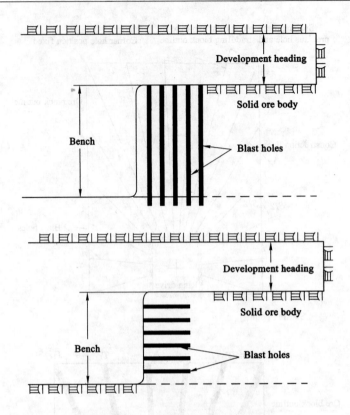

Figure 3-31 Long-hole bench blasting

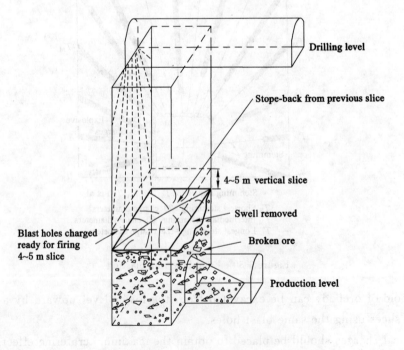

Figure 3-32 Long-hole VCR blasting

Chapter 3 Rock excavation methods

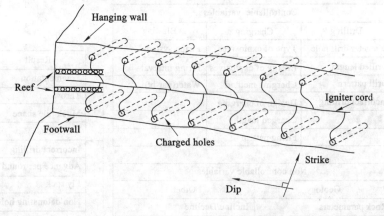

Figure 3-33 Short-hole blasting

The diameter and length of shot holes are usually limited to 43 mm and 4 m respectively.

Short-hole blasting is usually used in breast stoping for narrow, tabular orebodies such as gold or platinum reefs.

(3) Drill and blast: fragmentation(as shown in Figure 3-34)

Figure 3-34 Fragmentation

How effectively muck from a working face can be removed is a function of the blast fragmentation.

Broken rock by volume is usually 50% greater the in situ material.

In mining, both the waste and the ore have to be moved to the surface for milling or disposal.

3.2.9 Drill and blast: summary(as shown in Figure 3-35)

This section introduces in detail the common rock drilling and blasting methods, drill bit forms, blasting methods, drilling length and other issues, for readers to understand the rock excavation is of great significance.

In the mining process, a large number of underground roadways must be excavated to

Controllable variables		
Drilling	Charging	Blasting
Diameter drill hole	Type of explosives	Firing system
Drilled length	Energy of explosives	Firing interval
Drill pattern	Charging method	Water(partly)
Incorrect drilling	Design of charging	
	Charged length	
	Firing pattern	

Non controllable variables	
Geology	Other
Rock parameters	Incline/Decline
Rock mass joints	Water(partly)

Result
Fragmentation
Throw
Muck pile shape
Vibrations
Incorrect drilling
Advance per round
Fly rock
Non detonating holes
Poor blast results

Figure 3-35 Drill and blast: summary

assist the production of ore, ore excavation, transportation, and roof support need to excavate a certain workspace to meet the production requirements. With the development of coal mining in China, more and more problems are restricting the construction of underground roadways. Therefore, it is urgent to study and solve the rapid construction method of roadways under high stress and complex geological environment. This section will carry out the method of rock excavation in this section.

From the early days, the method developed from manual drilling of drill bits and hammering holes and the detonator detonating individual charge packets one by one, to drilling with the drilling rig or multi-arm drilling vehicle, and the blasting techniques such as millisecond blasting, pre-splitting blasting and smooth blasting were applied. Before construction, the method of tunneling shall be selected according to geological conditions, section size, support method, construction period requirements, construction equipment, technology and other conditions. Borehole blasting is always the main method of rock excavation in underground buildings. This method has strong adaptability to geological conditions and low cost of excavation, and is especially suitable for the construction of hard rock cavities. The blasting design should be made according to the design requirements, geological conditions, blasting materials and drilling equipment.

3.3 Blasting damage in rock

3.3.1 Blasting damage in rock (Hoek, 2007)

The impact of blasting damage upon the stability of structures in rock is not widely recognised or understood.

The precise nature of the mechanism of rock fragmentation as a result of blasting is not fully understood. However, from a practical point of view, it is safe to assume that both the dynamic stresses induced by the detonation and the expanding gases produced by the explosion play important roles in the fragmentation process.

Influence of geological structures[63-66]:

① The strength of jointed rock masses is controlled by the degree of interlocking between individual rock blocks separated by discontinuities such as bedding planes and joints.

② For practical purposes, the tensile strength of discontinuities can be taken as zero, therefore a small amount of opening or shear displacement will result in a dramatic drop in the interlocking of the individual blocks.

High pressure gases expanding outwards from an explosion will likely jet into the discontinuities and cause a breakdown of the natural block interlocking. Controlling factors include:

① The distance from the explosive charge.

② The in situ stresses, which have to be overcome by the high pressure gases before loosening of the rock can take place.

The extent of the gas pressure induced damage can be expected to decrease with depth below surface, and surface structures such as slopes will be very susceptible to gas pressure induced blast damage.

An additional cause of blast damage is that of fracturing induced by release of load (Hagan, 1982).

Think of…dropping a heavy steel plate onto a pile of rubber mats. These rubber mats are compressed until the momentum of the falling steel plate has been exhausted. The highly compressed rubber mats then accelerate the plate in the opposite direction and, in ejecting it vertically upwards, separate from each other. Such separation between adjacent layers explains the "tension fractures" frequently observed in open pit and strip mine operations where poor blasting practices encourage pit wall instability.

McIntyre and Hagan (1976) report vertical cracks parallel to and up to 55 m behind newly created open pit mine faces where large multi-row blasts have been used.

Cracks can be induced at very considerable distance from the point of detonation of an explosive. These fractures can have a major disruptive effect upon the integrity of the rock mass→reduction in overall stability[67].

Impact on slope stability:

① Hoek (1975) has argued that blasting will not induce deep seated instability in large open pit mine slopes.

② Failure surface can be several hundred metres below the surface in a very large slope, and the failure surface is generally not aligned in the same direction as blast induced fractures.

③ But…near surface damage to the rock mass can seriously reduce the stability of the individual benches which make up the slope and which carry the haul roads.

④ In a badly blasted slope, the overall slope may be reasonably stable, but the face may resemble a rubble pile.

When observed from a distance, the overall pit slope may look reasonably stable, but look closer⋯and the face may resemble a rubble pile. As shown in Figure 3-36.

(a)　　　　　　　　　　　　(b)

Figure 3-36　Example diagram

Impact on slope stability (continued):

A common misconception is that the only step required to control blasting damage is to introduce pre-splitting or smooth blasting techniques.

Pre-splitting = simultaneous detonation of a row of closely spaced, lightly charged holes, designed to create a clean separation surface between the rock to be blasted and the rock which is to remain.

Controlling blasting damage must start long before the introduction of pre-splitting or smooth blasting.

No undo button⋯If blast damage has already been inflicted on the rock, it is far too late to attempt to remedy the situation by using smooth blasting to trim the last few metres of excavation[68].

Spot the difference⋯

Comparison between the results achieved by pre-split blasting (on the left) and normal bulk blasting for a surface excavation in gneiss (on the right). As shown in Figure 3-37.

Impact on tunnel stability:

In a tunnel or other large underground excavation, the stability of the underground structure is very much dependent upon the integrity of the rock immediately surrounding the excavation[69].

The tendency for roof falls is directly related to the interlocking of the immediate roof strata.

Since blast damage can easily extend several metres into the rock which has been poorly blasted, the halo of loosened rock can give rise to serious instability problems in the

Chapter 3   Rock excavation methods

Figure 3-37   Comparison between the results achieved by pre-split blasting (on the left) and normal bulk blasting for a surface excavation in gneiss (on the right)

rock surrounding the underground openings.

Damage control:

The ultimate in damage control is machine excavation (→to be discussed later as part of this lecture).

Think about a bored raise and the general lack of disturbance associated with it.

Even when the stresses in the rock surrounding the raise are high enough to induce fracturing in the walls, the damage is usually limited to less than half a metre in depth, and the overall stability of the raise is seldom jeopardised.

The machine excavation techniques are not widely applicable in all underground mining situations, and therefore consideration must be given to what can be done about controlling damage in normal drill and blast operations[70,71].

In summary:

Damage is being caused to both tunnels and surface excavation by poor blasting. This damage results in a decrease in stability which, in turn, adds to the costs of a project by the requirement of greater volumes of excavation or increased rock support.

In general terms many of the existing techniques, if correctly applied, could be used to reduce blasting damage in both surface and underground rock excavation.

Poor communications and reluctance to become involved on the part of most engineers, means that good blasting practices are generally not used on mining and civil engineering projects.

Need for a clear set of principles on blasting design and control, written in unambiguous, non-mathematical language.

A well designed blast is generally more efficient and may provide improve fragmentation and better muck-pile conditions at the same cost.

3.3.2 Assessment of blast damage (Singh and Lamond, 1994)

To assess the damage created by blasting the following methods can be used.

① Vibration monitoring.
② Visual conditions.
③ Percentage over break.
④ Half cast factor.

Vibration monitoring is one of the most important tools in understanding blasting. For example, complete seismic records of each blast could be recorded and subsequently analysed.

Visual condition of the remaining rock wall and back (roof) is a crude but sometimes a reliable method of determining blast damage.

Quantifying tools such as half cast factor and % over break are more reliable means.

Percentage over break is a measure of the amount of rock that is removed from the periphery of an excavation beyond the planned limit. Percentage over break is determined by comparing the designed profile with the after blast profile.

Half cast factor is a measure of the remaining half casts (half barrels) left on the rock walls and back. However, half cast factor assessment does not take into account any error of borehole alignment.

If the drill hole alignment has deviated from the designed pattern, break measurements will calculate this as damage. But using the half cast factor assessment, it is possible to determine that damage was created by drilling error rather that from blasting.

3.4 Mechanical excavation in rock

We have already introduced the concept of mechanical excavation in rock when discussing damage control for blasting.

Mechanical excavation methods:

① Are a step change in excavation performance and labour safety compared to drill and blast operations.

② Are becoming more and more common, particularly for civil engineering tunnelling.

③ Are capable of increasing overall mine performance because more process steps can take place simultaneously.

Machines have been developed to the point where advance rates and overall costs are generally comparable or better than the best drill and blast excavation methods.

The lack of disturbance to the rock and the decrease in the amount of support required are major advantages in the use of tunnelling machines.

The machine excavation techniques described are not widely applicable in all underground mining situations, and therefore consideration must be given to what can be done about controlling damage in normal drill and blast operations.

There are two basic types of machine for underground rock excavation:

① Partial-face machines: use a cutting head on the end of a movable boom, generally track mounted (→road header machines).

② Full-face machines: use a rotating head armed with disc cutters, which fills the tunnel cross section completely, and thus almost always excavates circular tunnels (→ TBM or tunnel boring machines). As shown in Figure 3-38.

Figure 3-38 Mechanical excavation equipment

Two main types of mechanical excavation tools are used:

Cutter bits (or picks)→high impact loads under a low angle to the rock surface.

Disc cutters→ rolling over the surface and penetrating the rock in a perpendicular direction.

The advantages of cutter bits include lower cutting forces and more flexible cutting kinematics (→ small and lightweight equipment suitable for a more flexible excavation geometry).

Once the rock strength exceeds the range of 100~120 MPa unconfined compressive strength, rock excavation with cutter bits becomes increasingly infeasible due to low penetration rates and high bit consumption.

Conversely, disc cutters are able to excavate rock with a compressive strength of more than 300 MPa, but require significantly higher cutting forces which lead to heavier and less mobile equipment.

To achieve high excavation rates, disc cutters are usually employed on rotary full-face TBM.

3.4.1 Road header machines

(1) Road header machines, although slower than tunnel boring machines, are capable of excavating more complex geometries and arc used as an alternative to drill & blast to develop auxiliary openings leading from the main TBM tunnel.

he cutter head swings up and down and across the face of the heading, cutting the rock and dropping rock fragments onto a steel apron. Gathering arms then move the muck onto a conveyor and out of the excavation. As shown in Figure 3-39.

(2) TBM machines:

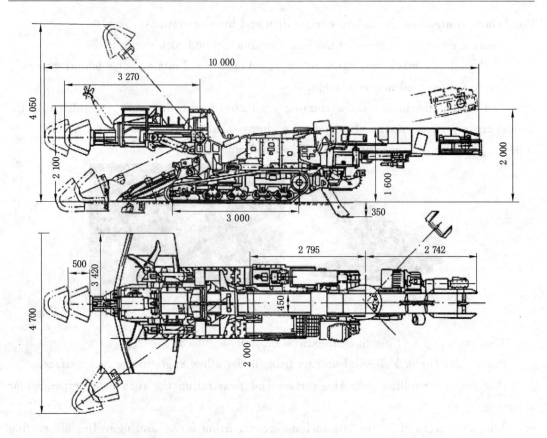

Figure 3-39 Road header machines

In hard rock conditions, road headers cannot be used effectively, so disc cutting is the first choice.

Single and double shield TBMs:

Single shield TBMs are cheaper and are proffered for hard rock tunnelling.

Double shield TBMs are normally used where difficult ground conditions are expected (as they offer more protection for the workers) or when a high rate of advancement is required. As shown in Figure 3-40.

The advance rate at which the excavation proceeds is a function of the cutting rate

Chapter 3　Rock excavation methods

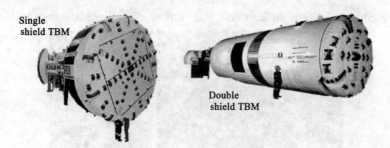

Figure 3-40　Single and double shield TBMs

andutilization factor (time that the machine is actually cutting the rock).

Factor contributing to low utilization rates are:

Difficult ground conditions (ground support requirements).

Need to replace cutters.

Blocked conveyors.

Blocked scoops.

(3) Mechanics of rock cutting(as shown in Figure 3-41):

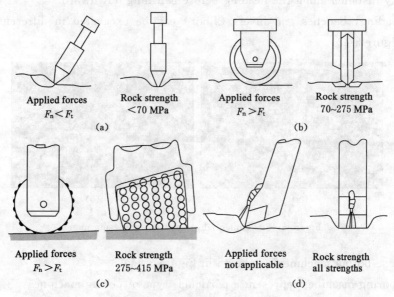

Figure 3-41　Rock cutting stress diagram
(a) Drag pick;(b) Disc cutter;(c) Button cutter;(d) Jet-assisted cutter

When considering using a TBM, both applied thrust (F_n) and torque (F_t) have to be considered.

The engineer should think about how configure the cutting tools in order to:

Minimize the need to replace the cutters.

Avoid damaging the cutter mounts.

Minimise vibration.

Replacing cutters can be a difficult and time consuming task⋯(as shown in Figure 3-42).

Figure 3-42　Rock cutting map

(4) Sequential excavation and design:

Sequential excavation by benches is typically used for large diameter excavations in weak rock.

Easier to control ground conditions with a small opening and reinforcement can be progressively installed along the heading before benching downward.

Top heading, benches and invert (floor) can be excavated in different orders (as shown in Figure 3-43).

Figure 3-43　Sequential excavation and design map

(5) Raise boring machines(as shown in Figure 3-44):

Raise boring machines represent a particular type of boring machines:

Used for excavating ventilation shafts and ore passes for mines.

Also used for infrastructural projects for tunnel ventilation or surge shaft in hydro-electric power projects.

The raise borer is generally set up on a concrete pad on the upper level of the two levels to be connected. A small-diameter hole (pilot hole) is initially drilled to the levelrequired 230～350 mm diameter. Once the drill has broken into the opening on the target level, the bit is removed and a reamer head, of the required diameter of the excavation, is attached to the drill string and raised back towards the machine.

Largest diameter raise of 7.1 m and longest excavation 1 260 m.

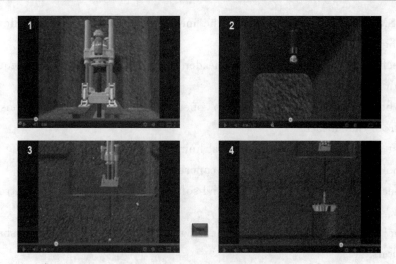

Figure 3-44   Raise boring machines

3.4.2 Shaft boring systems

Fast access to the orebody is critical in any mining methods, and block caving in particular. Excavation of shafts or declines is typically a critical path of the project schedule (→depth of the orebody, undercut and production levels).

If we can save time (but remember that safety always comes first) on the excavation of shafts or declines, then we can significantly increase the net present value of the mining project.

For deep hard rock shafts, shaft boring system (SBS) are now available, integrating excavation, mucking, primary rock support, installation of the final lining, and shaft infrastructure.

This new system dramatically improves the health and safety of shaft construction.

(1) Shaft boring systems: vertical shaft machine (VSM)(as shown in Figure 3-45):

Figure 3-45   Vertical shaft machine (VSM)

The VSM has been developed for the mechanised excavation of shallow shafts in water-bearing soils.

This technology is based on a roadheader boom with a cutter drum equipped with cutter bits or soft ground chisels.

The whole machine is designed to operate in submerged conditions, remotely controlled from the surface.

Pre-cast concrete segments (segmental lining).

Rock bolts, wire mesh, and sprayed concrete.

VSM technology can be used in soil and soft rock conditions with rock strengths of up to 120 MPa.

(2) Shaft boring system (SBS) for excavation of deep vertical blind shafts in hard rock conditions.

Rotating cutting wheel excavating the full shaft diameter in a two stage process for one complete stroke.

Trench excavation to a depth of one stroke with the cutting wheel rotating around its horizontal axis and being pushed downward in the shaft direction.

Excavation of the entire bench (face) area by slewing the rotating cutting wheel 180° around the shaft vertical axis(as shown in Figure 3-46).

Figure 3-46　Shaft boring systems

The SBS machine can be separated into the main functional areas (starting from the bottom):

① Excavation chamber with cutting wheel, cutting wheel drive assembly, mechanical machine support structure, shotcrete and probe drilling equipment.

② Adjustable front support with slew bearing/drive assembly cutting wheel support and dust shield.

③ Regular rock support area for rock bolts.

④ SBS mainframe with gripper carrier, gripper system and thrust cylinders.

⑤ Rear alignment system (secondary gripper) and muck handling system.

During the gripper reset operation after each excavation cycle, the machine can be adjusted along its vertical axis for alignment control(as shown in Figure 3-47).

Figure 3-47　SBS main functional zoning

Chapter 4 Rock excavation and support

The wide variety of orebody shapes and rock mass characteristics which are encountered in underground mining mean that each mine presents a unique design challenge.

There is no typical mining method and conditions may have to be modified to fit the site-specific conditions.

Needles to say, any chosen system must comply with safety and stability requirements. Figure 4-1 shows the support form for different roadway sections.

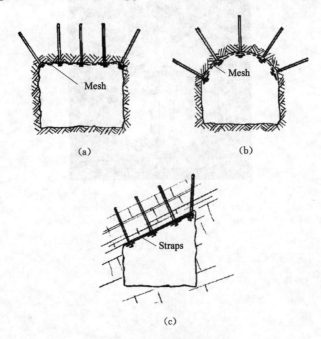

Figure 4-1 Support of different roadway sections
(a) Conventional rectangular excavation; (b) Arched roof excavation; (c) Shanty back excavation in bedded rock

4.1 Bolts, dowels and cables

We can expect safety bolts or dowels generally not requiring to carry a load in excess of about one ton.

Mechanically anchored rock bolts or friction anchored dowels (e. g. swellex or split sets) are adequate for these installations.

How should we choose which system to use?

Costs and availability, ease and speed of installation.

Corrosion problems (→protective coating or grouted in place).

Rock mass blockiness (very closely jointed rock mass) and need to prevent small blocks and wedges falling out between the rock bolts.

Wire mesh, chain link mesh installed behind the rock bolt washers or face plates.

Mesh and corrosion problems use of steel fibre reinforced shotcrete instead.

Straps can also be useful for providing support between rock bolts installed in bedded rock masses… but straps installed parallel to the strike of significant discontinuities in the rock will serve little purpose.

Shafts, shaft stations and underground crusher chambers are examples of permanent mining excavations (high usage, high capital costs).

These excavations are typically designed for an operational life of tens of years.

Therefore corrosion is a problem which cannot be ignored.

Galvanised or stainless steel rock bolts.

Fully grouted dowels, rock bolts or cables (usually more effective and economical).

Use of fibre or mesh-reinforced shotcrete, rather than mesh or straps, on exposed surfaces. The flow chart of support mode selection is shown in Figure 4-2.

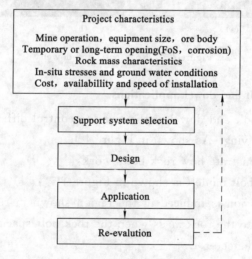

Figure 4-2 Support selection process

Rock bolts and dowels have been used for many years for the support of underground excavations and a wide variety of bolt and dowel types are available to meet different demands in mining and civil engineering.

Rock bolts generally consist of plain steel rods with a mechanical or chemical anchor at one end and a face plate and nut at the other.

They are always tensioned after installation.

For short term applications the bolts are generally left ungrouted, for more permanent applications (or to protect against corrosion), cement or resin grout is used.

Dowels (anchor bars) generally consist of deformed steel bars which are grouted into the rock.

Tensioning is not possible and the load in the dowels is generated by movements in the rock mass.

Dowels have to be installed before significant movement in the rock mass has taken place.

Cables represent a type of reinforcement technology used to take on the support duties which exceed the capacity of traditional rock bolts and dowels. Bolting and pallet support are shown in Figure 4-3.

(a)　　　　　　　　(b)

Figure 4-3　Rock bolts and dowels

4.1.1　Typical rock bolt and dowel applications

(1) Typical rock bolt and dowel applications to control different types of rock mass failure during tunnel driving. As shown in Figure 4-4.

(2) Model to demonstrate how rock bolts work:

Tom Lang's rock bolt model (as shown in Figure 4-5):

A zone of compression is induced in the region shown in red and this will provide effective reinforcement to the rock mass when the rock bolt spacing s is less than 3 times the average rock piece diameter a.

The rock bolt length L should be approximately $2s$. Note there is no support between the washers (unless mesh or shotcrete is applied) and the rock pieces will fall out of these zones on the underside of the beam.

Model with temporary base removed from the self-supporting rock plate. As shown in Figure 4-6.

Demonstration of the load-carrying capacity of the bolted gravel plate.

Note the rock pieces fallen out of the zones on the underside of the beam in between the rock bolts. The experiment is shown in Figure 4-7.

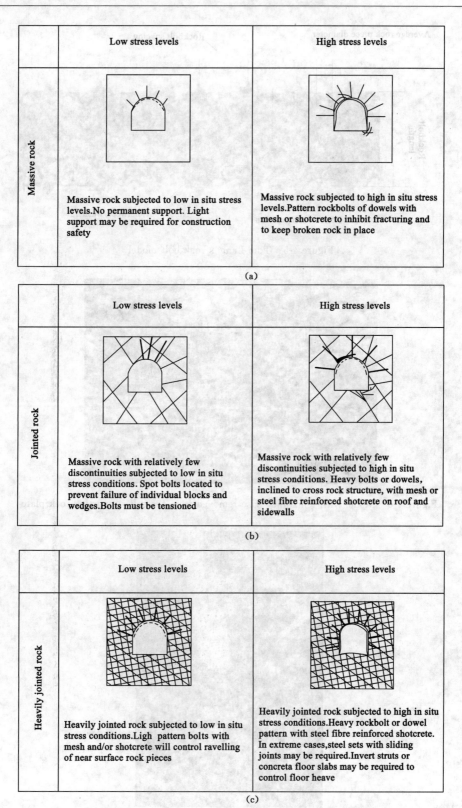

Figure 4-4 Typical rock bolt and dowel applications

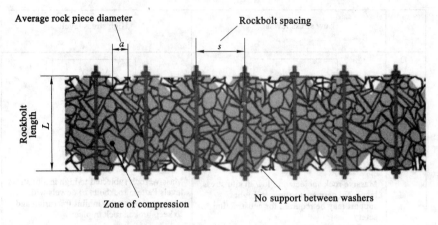

Figure 4-5　Tom Lang's rock bolt model

Figure 4-6　Model with temporary base removed from the self-supporting rock plate

Figure 4-7　Field test of rock bolt anchorage

Chapter 4   Rock excavation and support

4.1.2 Types of anchor bolts

(1) Mechanically anchored rock bolts(as shown in Figure 4-8 and Figure 4-9)

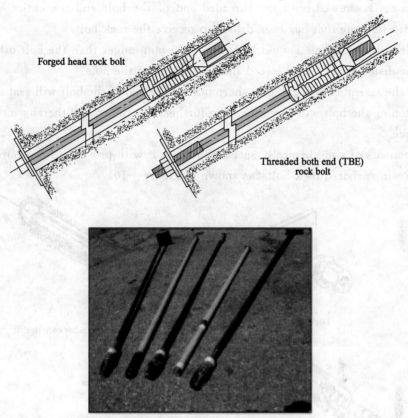

Figure 4-8   Structure and action mechanism of mechanical anchored bolt

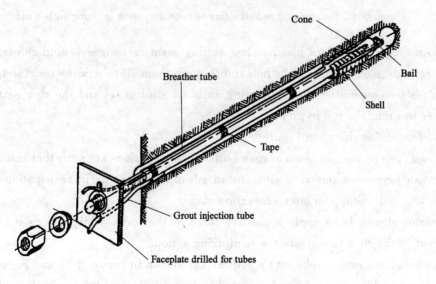

Figure 4-9   Composition of mechanical anchored bolt

· 81 ·

The components of a typical expansion shell anchor area tapered cone with an internal thread and a pair of wedges held in place by a bail.

The cone is screwed onto the threaded end of the bolt and the entire assembly is inserted into the hole that has been drilled to receive the rock bolt.

The length of the hole should be at least 100 mm longer than the bolt otherwise the bail will be dislodged by being forced against the end of the hole.

Once the assembly is in place, a sharp pull on the end of the bolt will seat the anchor.

Tightening the bolt will force the cone further into the wedge thereby increasing the anchor force.

Components of a mechanically anchored rock bolt with provision for grouting.

(2) Resin anchored rock bolts(as shown in Figure 4-10)

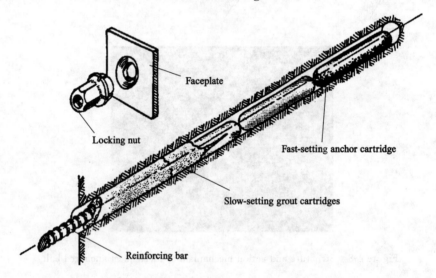

Figure 4-10 Composition and structure of resin anchoring grouting anchor rod

Resin grouting involves placing slow setting resin cartridges behind the fast-setting anchor cartridges and spinning the bolt rod through them all to mix the resin and catalyst. The bolt is tensioned after the fast-setting anchor resin has set and the slow-setting resin sets later to grout the rod in place.

(3) Dowels(as shown in Figure 4-11)

Dowels can be used in place of rock bolts when conditions are such that installation of support can be carried out very close to an advancing face, or in anticipation of stress changes that will occur at a later excavation stage.

Tensioned rock bolts apply a positive force to the rock, while dowels depend upon movement in the rock to activate the reinforcing action.

(4) Friction dowels (split set) stabilizers(as shown in Figure 4-12 and Figure 4-13)

Split set stabilizers consists of a slotted high strength steel tube and a face plate. It is installed by pushing it into a slightly undersized hole and the radial spring force generated,

Chapter 4 Rock excavation and support

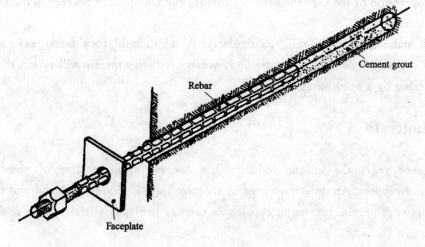

Figure 4-11 Dowels

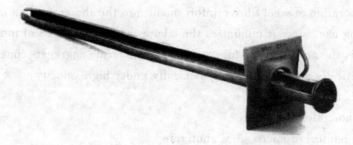

Figure 4-12 Friction dowels (split set) stabilizers

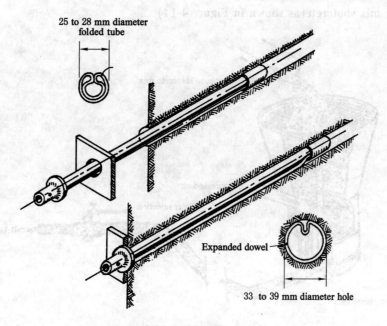

Figure 4-13 Swelled system (atlas copco)

by the compression of the C shaped tube, provides the frictional anchorage along the entire length of the hole.

Quick and simple to install. Particularly useful in mild rock burst environments, because it will slip rather than rupture and, when used with mesh, will retain the broken rock generated by a mild burst.

4.2 Shotcrete

In recent years the mining industry has become a major user of shotcrete for underground support. An important area of shotcrete application in underground mining is in the support of permanent openings such as ramps, haulage, shaft stations and crusher chambers.

Rehabilitation of conventional rock bolt and mesh support can be very disruptive and expensive.

The incorporation of steel fibre reinforcement into the shotcrete is an important factor in this escalating use, since it minimizes the labour intensive process of mesh installation.

Shotcrete is the generic name for cement, sand and fine aggregate concretes which are applied pneumatically and compacted dynamically under high velocity:

Dry mix shotcrete.

Wet mix shotcrete.

Steel fibre reinforced micro silica shotcrete.

Mesh reinforced shotcrete.

4.2.1 Dry mix shotcrete (as shown in Figure 4-14)

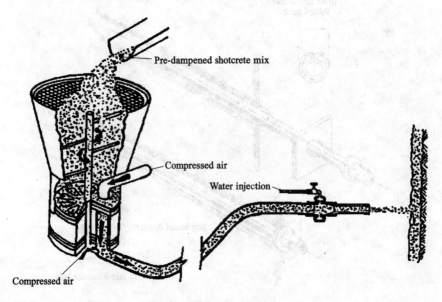

Figure 4-14 Dry mix shotcrete

The dry mix system tends to be more widely used in mining, because of inaccessibility for large transit mix trucks and because it generally uses smaller and more compact equipment.

4.2.2 Steel fibre reinforced micro silica shotcrete(as shown in Figure 4-15)

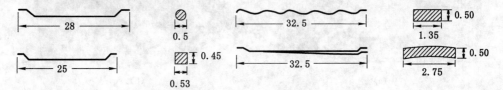

Figure 4-15　Steel fibre reinforced micro silica shotcrete

Use of silica fume, as a cementitious admixture, and steel fibre reinforcement.

Silica fume or micro silica is a by-product of the ferro silicon metal industry and is an extremely fine pozzolan (pozzolans are cementitious materials which react with the calcium hydroxide produced during cement hydration).

Silica fume, added in quantities of 8% to 13% by weight of cement, can allow shotcrete to achieve compressive strengths which are double or triple the value of plain shotcrete mixes.

Steel fibre reinforced shotcrete was introduced in the 1970s.

The main role that steel fibre reinforcement plays in shotcrete is to impart ductility to an otherwise brittle material.

4.3　Support design

Both empirical and numerical methods can be used for support design for underground openings.

The magnitude of support is influenced by the rock mass characteristics, in-situ stresses, and the excavation geometry.

Barton's Q system provides a chart for tunnel support based on the Q rating, span and use of the excavation.

Numerical codes (e.g. Phase 2) provide a convenient means of modeling the rock mass mechanical interaction of the ground and support by staging the project according to the excavation sequence. As shown in Figure 4-16.

4.3.1　Failure mechanism

It is important to distinguish between kinematic analysis and stress analysis.

For the first, the bolt pattern is designed to support wedges that may form due to discontinuities intersecting with the excavation. In the stress analysis, the model attempts to predict the overall response of the rock mass in terms of stress and deformation. As shown in Figure 4-17.

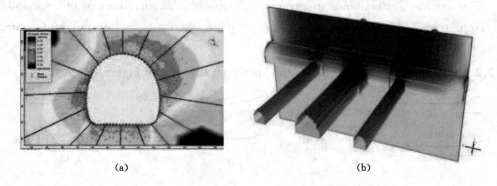

Figure 4-16 Numerical simulation image

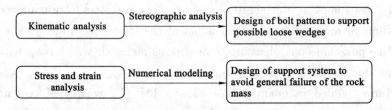

Figure 4-17 Analysis of support design

4.3.2 Capacity of grouted dowel(as shown in Figure 4-18)

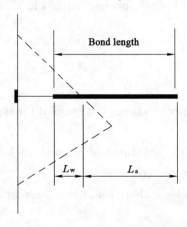

Figure 4-18 Capacity of grouted dowel

(1) For grouted dowels there are three different failure mechanisms: pullout (force required to pull the anchorage length L_a of the bolt out of the rock), tensile failure, and stripping(wedge failure occurs, and bolt remains embedded in rock).

Dowel parameters:

T = Tensile capacity (force);

P = Plate capacity (force);

B = Bond strength (force/unit length of bond).

The maximum force which can be mobilized by each failure mode, at any point along the bolt length, is given by the following equations:

Pullout: $\quad F_1 = BL_a$

Tensile: $\quad F_2 = T$

Stripping: $\quad F_3 = P + BL_w$

The limiting failure mechanism is given by the minimum of these three forces.

(2) Rock support interaction:

Following the excavation, the rock mass gradually deforms. In turn, the pressure required to hold the rock in place decreases.

Weaker rock masses will exhibit larger deformations and will therefore convey a larger pressure to a given support system compared to stronger rock masses.

Enabling some rock deformation to occur prior to support installation may increase the FoS or alternatively permit the use of a support with a lesser capacity. However, displacements that are too large (i.e. cause stresses to exceed the plastic limit) can result in collapse. As shown in Figure 4-19.

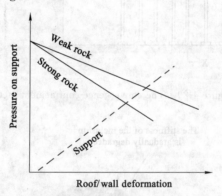

Figure 4-19   Variation diagram

(3) The convergence-confinement method (as shown in Figure 4-20):

Essentially, installing support during tunnel excavation is a 4D problem as it is influenced by the timing of the installation and the distance of the support installation to the tunnel face[73].

Modeling the problem in 2D may therefore yield over-conservative results as both the support of the face and the release of pressures due to initial deformations are neglected.

In the convergence confinement method reaction curves for the support and ground are developed and their intersection represents the final rock displacement and support pressures.

This enables optimization of the distance and timing of support installation.

(4) Core replacement technique (as shown in Figure 4-21):

The core replacement technique proposed by Hoek enables simulating the 4D effects with a 2D analysis by gradually degrading the stiffness of the material enclosed in the

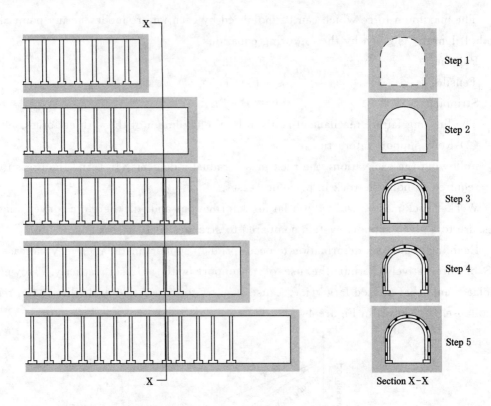

Figure 4-20   The convergence-confinement method

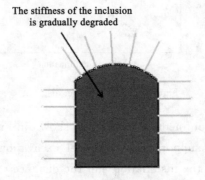

Figure 4-21   Core replacement technique

tunnel throughout the project stages.

The displacement that occurs at a given distance from the tunnel face can be found using empirical formulae.

The simulated support must be assigned to the stage with an inclusion that corresponds to the above displacement.

(5) Reinforced shotcrete liner design:

The external forces imposed by the rock mass on the liner are in the form of bending moments, shear forces and thrust forces.

To satisfy force equilibrium, internal stresses develop in the shotcrete.

The factor of safety can be defined as the strength of the shotcrete divided by the stresses that develop as a result of the external forces.

In the design process, if the FoS is not sufficient, the thickness of the shotcrete can be increased or alternatively reinforcement can be added. As shown in Figure 4-22.

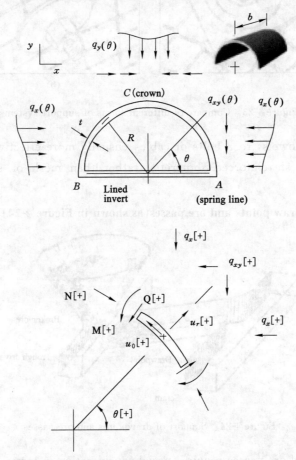

Figure 4-22 Reinforced shotcrete liner design

(6) Different forms of support systems(as shown in Figure 4-23):

Wire mesh should be firmly attached to the rock by washers or face plates on the rock bolts or dowels.

Welded wire mesh isa better choice than chain link mesh where the excavation surfaces are reasonably smooth and where there is enough room to work.

Shafts, shaft stations and underground crusher chambers are examples of permanent mining excavations (high usage, high capital costs).

These excavations are typically designed for an operational life of tens of years. Therefore corrosion is a problem which cannot be ignored. galvanized or stainless steel rock bolts.

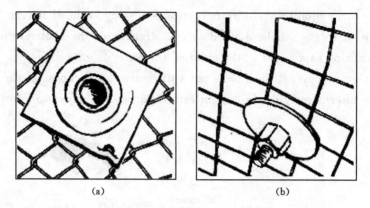

Figure 4-23 Compare of different forms of support systems

Fully grouted dowels, rock bolts or cables (usually more effective and economical). Use of fibre or mesh-reinforced shotcrete, rather than mesh or straps, on exposed surfaces.

4.3.3 Support of draw points and ore passes(as shown in Figure 4-24)

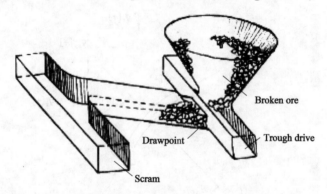

Figure 4-24 Support of drawpoints and ore passes

Draw points and ore passes require special consideration in terms of support design.

These openings are generally excavated in undisturbed rock, therefore mining is relatively easy and little support is required to stabilize the openings, however… once mining starts and the draw points and ore passes are brought into operation, the conditions are changed dramatically and serious instability can occur if support has not been installed in anticipation of these changes.

Abrasion (passage of hundreds of tonnes of broken ore) and raveling.

Stress changes due to the mining of adjacent or overlying stopes and caves.

Secondary blasting of hang-ups in the draw points or ore passes.

There is considerable economic incentive to install the correct reinforcement during development of the openings in order to avoid costly remedial work later.

Plain rebar's, with no face plates or end fixings, should be used to facilitate the movement of the ore through the draw point and to avoid the face plates from being ripped off.

Surface coatings such as shotcrete should only be used where the surrounding rock is clean and of high quality and where the draw point is only expected to perform light duty.

Grouted rebar is a good choice for draw point reinforcement in cases where the rock is hard, strong and massive.

When the rock is closely jointed and there is the possibility of a considerable amount of inter-block movement during operation of the draw point, rebar may be too stiff and the rock may break away around the rebar → consider grouted birdcage or uncage cables instead.

The design of support for ore passes is similar to that for draw points, but on ore pass is required to handle much larger tonnages of ore and may be required to remain in operation for many years.

Access to install the support is generally not as simple as for draw points. Identification of weak zones and provision of adequate reinforcement during construction.

Support, where possible, should be installed from inside the ore pass during excavation.

Intentioned, fully grouted birdcage cables are probably the best type of reinforcement, since they have a high load carrying capacity for their whole length and he projecting ends will not obstruct the passage of the ore.

Example of suggested reinforcement for a draw point in a large mechanized mine. As shown in Figure 4-25.

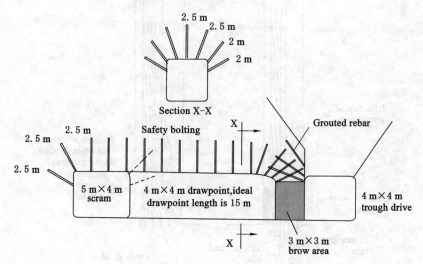

Figure 4-25   Support of draw points and ore passes(Hoek, Kaiser and Bawden, 1995)

The brow area, shown shaded, is blasted last after the rebar has been grouted in place from the draw point and trough drive. Safety bolting can be used in the draw point and scram.

One of the primary causes of rock bolt failure is rusting or corrosion and this can be counteracted by filling the gap between the bolt and the drill hole wall with grout.

While this is not required in temporary support applications, grouting should be considered where the ground-water is likely to induce corrosion or where the bolts are required to perform a permanent support function.

Many of the difficulties associated with installation of grouted bolts can be overcome by using a hollow core bolt.

The primary purpose of grouting mechanically anchored bolts is to prevent corrosion and to lock the mechanical anchor in place, therefore the strength requirement for the grout is not as important as it is in the case of grouted dowels or cables.

The grout should be readily pump able without being too fluid and a typical water/cement ratio of 0.4 to 0.5 is a good starting point for a grout mix for this application.

4.3.4 Up hole (breather tube method) installation and grouting (as shown in Figure 4-26)

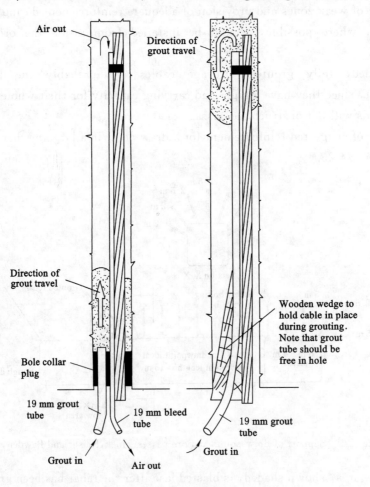

Figure 4-26 Alternative methods for grouting cables in upholes

The grout, usually having a water/cement ratio $\geqslant 0.4$, is injected into the bottom of the hole through a large diameter tube, typically 19 mm diameter.

The air is bled through a smaller diameter tube which extends to the end of the hole and which is taped onto the cable.

Both tubes and the cable are sealed into the bottom of the hole by means of a plug of cotton waste or of quick setting mortar. The direction of grout travel is upwards in the hole, therefore the grout column would be devoid of air gaps since any slump in the grout would tend to fill these gaps.

Apart from the difficulty of sealing the collar of the hole, the main problem with this system is that it is difficult to detect when the hole is full of grout.

In this case a large diameter grout injection tube extends to the end of the hole and is taped onto the cable.

Provided that a very viscous mix is used (0.3 to 0.35 water/cement ratio), the grout will have to be pumped into the hole and there is little danger of slump voids being formed.

However, a higher water/cement ratio mix will almost certainly result in air voids in the grout column as a result of slumping of the grout.

The principal advantage of this method is that it is fairly obvious when the hole is full of grout and this, together with the smaller number of components required, makes the method attractive when compared with the traditional method for grouting plain strand cables.

The thicker grout used in this method is not likely to flow into fractures in the rock, preferring instead the path of least flow resistance towards the borehole collar.

Chapter 5   Rock engineering design

The design and analysis of underground excavations can be discussed in the context of the same mechanisms we introduced for surface excavation design. As shown in Figure 5-1.

Figure 5-1   Wedge failure and spalling(Esterhuizen, et al., 2011)

① Structurally-controlled instability mechanisms (e.g. wedge failure)
As shown in Figure 5-2.

Figure 5-2   Structurally-controlled failure(Esterhuizen, et al., 2011)

② Stress-controlled instability mechanisms (e.g. spalling)
The non-ideal nature of rock in-situ (rock mass) as an engineer material has important implications in relation to engineering risk.

For example: what would be the effect of excavating a tunnel in the fractured rock mass below? As shown in Figure 5-3.

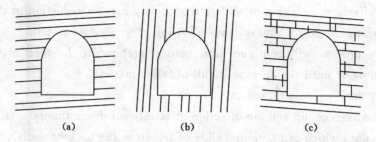

Figure 5-3 Fractured rock mass
(a) Sub horizontal strata; (b) Sub vertical strata; (c) Horizontal strata with vertical conjugate fractures

In principle, excavation design in massive (elastic) rock represents the simplest design problem posed in mining rock mechanics.

However, we need to take into consideration key questions related to rock strength, fracture and failure and their effect on the behavior of rock in the boundary and near field of the excavation.

There is the need to consider that two compressive strength criteria (S-shaped failure criteria) arises because different failure modes apply near the excavation boundary and in the interior of the rock.

① Under the complex stress path and in the low confinement conditions near the boundary of the excavation, crack initiation leads to unstable crack growth and the formation of spalls in the excavation boundary.

② Under confined conditions in the interior of the rock mass, rock failure depends on the formation of a population of interacting cracks, i.e. the accumulation of damage in the rock fabric from crack initiation and growth.

5.1 Structurally-controlled instability

The condition that arises in a blocky rock mass is the potential generation of discrete blocks, of various geometries defined by the natural discontinuities and the excavation surface.

Rock wedges are formed by intersecting discontinuities that separate the rock mass into discrete but interlocked pieces.

All blocks are formed by fractures but not all fractures form blocks.

A free surface is created when excavating the opening, and rock wedges can fall or slide from the surface if the bounding planes are continuous (or rock bridges along the discontinuities are broken).

Steps need to be taken to support potential unstable wedges, therefore ensuring the stability of the back (roof) and walls of the opening, which may otherwise rapidly deteriorate.

Domino's effect: each falling or sliding wedge may cause a reduction in the restraint and the interlocking of the rock mass, thus allowing other wedges to fall.

The failure process will continue until natural arching in the rock mass prevents further unravelling or until the opening is full of fallen material.

To prevent this, we would need to:

Determining average dip and dip direction of significant discontinuity sets.

Identify wedges which can potential slide or fall from the back or walls.

Calculate the factor of safety of these potentially unstable wedges, depending upon the mode of failure.

Calculate the amount of reinforcement required to bring the factor of safety of individual wedges up to an acceptable level.

As shown in Figure 5-4.

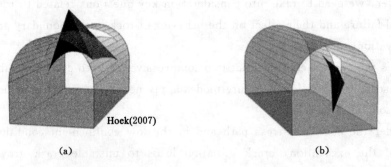

Figure 5-4 Two forms of structural instability
(a) Wedge falling from the roof; (b) Wedge sliding out of the sidewalls

5.1.1 Structurally-controlled instability and influence of stress

Stresses act as clamping forces acting on the surfaces of the potential wedges, and accordingly roof wedge stability can be significantly reduced when the in situ stresses are diminished for any reason.

It is very difficult to predict the in situ stresses precisely and to determine how these stresses can change with excavation of the tunnel or of adjacent tunnels or openings.

Typically, the design of the structure/underground excavation is carried out by neglecting the effect of stresses. This ensures that, for almost all cases, the support design will be conservative.

In rare cases the in situ stresses can actually result in a reduction of the factor of safety of sidewall wedges which maybe forced out of their sockets.

These cases are rare enough that they can generally be ignored for support design purposes.

5.1.2 Structurally-controlled instability & excavation sequence

(1) Rock wedges will fall or slide as soon as they are fully exposed in an excavated face, therefore requiring immediate support in order to ensure stability.

A support pattern can be designed and rock bolts can be installed as excavation progresses.

The problem of sequential support installation is typically presented when dealing with larger excavations.

It is important to define the structural geology of the surrounding rock mass to allow for a reasonable projection of potential wedges (design of loading capacity for the rock bolts).

These projections can be confirmed by additional mapping as each stage of the excavation is completed. As shown in Figure 5-5.

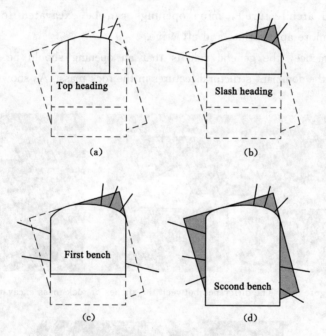

Figure 5-5 Structurally-controlled instability & excavation sequence

(2) Important considerations:

There are some important considerations which may have serious practical consequences if the effect of wedge size is ignored. A stage may inevitably be reached where the roof prism would collapse if it were decided to widen the opening.

Even a marginal increase in the span of an opening in a blocky rock mass can cause a significant reduction in the stability of the system, since it would correspond to a marked increase in the disturbing force (the block weight) relative to the mobilised resisting force. As shown in Figure 5-6.

Since mining engineering suffers from few of the cosmetic requirements of civil

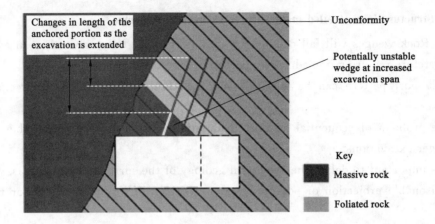

Figure 5-6 Effect of wedge size on stability(Brady and Brown, 2004)

engineering (and architecture), mine openings can be excavated to shapes that are geomechanically more appropriate and effective.

In mining practice, the general rule is that an opening should be mined to a shape conformable with the dominant structural features in the rock mass. As shown in Figure 5-7 and Figure 5-8.

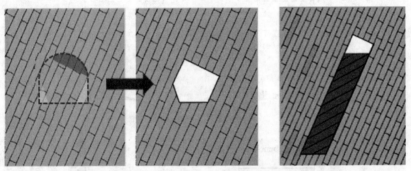

Figure 5-7 The difference between the shapes of rock mass excavation

Figure 5-8 Example: excavation shape controlled by structures

(3) A brief introduction to block theory:

① We have already discussed that the critical problem for engineering design in a blocky rock mass is the identification of the rock wedges that may be created and their stability state after the mining of a given excavation.

Block theory is the topological theory used to perform the identification of these rock wedges (Goodman and Shi, 1985).

The specific objective of block theory is to identify so-called key blocks or critical blocks that may pose a particular risk to the stability of an excavation.

The theory also accounts for the design of appropriate support and reinforcement, and for orientation of excavations and their boundaries to mitigate the effects of difficult block geometries.

② Block theory: important definitions.

The shapes and locations of key blocks are a fully 3D problem, however, the basic principles of block analysis can be understood from consideration of the 2D problem.

As an example, consider the 3 different types of blocks shown in Figure 5-9 and Figure 5-10:

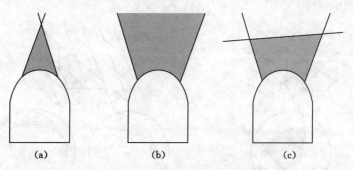

Figure 5-9　The there different types of blocks
(a) Finite and non-tapered;(b) Infinite;(c) Finite and tapered

Of these 3 block types, only the finite non-tapered block is kinematic ally capable of displacement into the excavation (→factor of safety against failure).

(4) Kinematic stability assessment:

Structurally controlled failure can be analysed by means of stereographic techniques, which provide a simple kinematic check that may be useful during preliminary studies. As shown in Figure 5-11.

Example of the use of stereographic projection to determine the shape and volume of a structurally defined wedge in the roof of a square tunnel:

Three great circles: A, B and C.

Strike lines: a, b and c.

Traces through the centre of the net and the great intersections: ab, ac and bc.

Assume a square tunnel with a spans in a direction 290°.

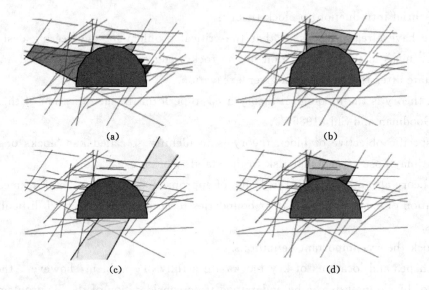

Figure 5-10 Example
(a) Finite and non-tapered blocks;(b) Composite block is finite and tapered;
(c) Infinite blocks;(d) Single blocks are finite and non-tapered

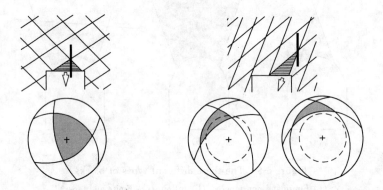

Figure 5-11 Condition for gravity fall of roof wedge and condition for sliding
failure of roof wedge unstable (left)-stable (right)

The directions of the strikes correspond to the traces of the planes A, B and C on the roof of the tunnel. These strike lines can be combined to give the maximum size of the triangular figure that can be accommodated within the tunnel.

We can define the height of the wedge by reading the angles α and β off the stereonet. As shown in Figure 5-12.

5.1.3 Stability assessment for underground excavations

(1) When considering the stability of underground excavations, the size and shape of potential wedges in the rock mass surrounding an opening is a function of:

Size, shape and orientation of the opening.

Length, frequency and orientation of the significant discontinuity sets.

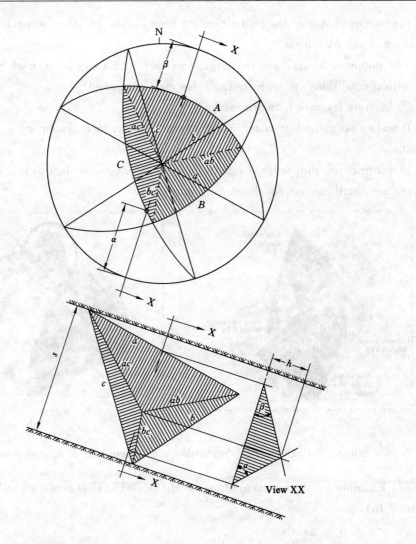

Figure 5-12 Example of the use of stereographic projection to determine the shape and volume of a structurally defined wedge in the roof of a square tunnel

Stereographic techniques are useful for gaining an understanding of structurally controlled failure and for checking the kinematic stability of isolated wedges in underground excavations.

Only really suitable for preliminary design.

Primarily evaluates critical orientations, neglecting other important joint properties.

When designing major excavations in well jointed rock masses, then various computer programs can be used for stability assessment. Different approaches are available, including:

Ubiquitous joints method (e.g. UNWEDGE).

Discrete fracture network method (e.g. FracMan).

① Ubiquitous joints method.

It assumes that the discontinuities are ubiquitous, in other words, they can occur anywhere in the rock mass.

The potential wedges are the largest wedges which can be formed for the assumed geometrical conditions (though scaling rules may be applied)

② Discrete fracture network method.

It makes no assumptions or limitations of the fracture system or its block forming potential.

Discontinuities properties (orientation, frequency and length) are described by appropriate distributions. As shown in Figure 5-13.

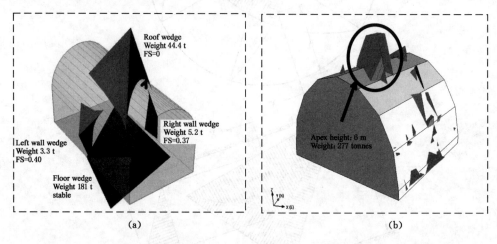

Figure 5-13 Ubiquitous joints method and discrete fracture network method

(2) Example 1: ubiquitous joints method (UNWEDGE)(as shown in Figure 5-14 and Figure 5-15):

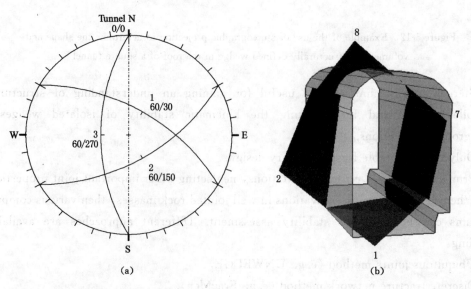

Figure 5-14 Three strongly developed joint sets occur (Rocscience, 2012)

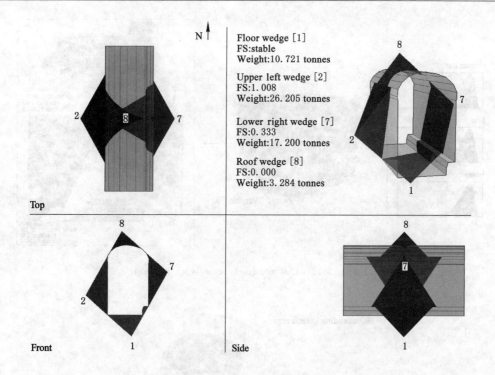

Figure 5-15 Four wedge figures

Consider a rock mass in which three strongly developed joint sets occur.

For the example under consideration, four wedges are formed.

All structural features included in the analysis are assumed to be planar and continuous.

These conditions mean that the analysis will always find the largest possible wedges.

This result can generally be considered conservative since the size of wedges, formed in actual rock masses, will be limited by the persistence and the spacing of the structural features.

These wedges are the largest wedges which can be formed for the given geometrical conditions.

(3) Example 2: discrete fracture network method(as shown in Figure 5-16):

This is a probabilistic method as it identifies blocks that are geometrically possible whilst also determining the likelihood of their formation.

(4) Example 3: discrete fracture network method:

Study of the stability/dilution problems that may be encountered at the contact between the host rock and a kimberlite pipe. As shown in Figure 5-17.

Rock block A: volume: 1 205 m^3, apex height: 13 m, FoS: 0.2.

Rock block B: volume: 928 m^3, apex height: 7.9 m, FoS: 0.15.

Rock block C: volume: 45 m^3, apex height: 3.5 m, FoS: 0.3. As shown in Figure 5-18.

Because this is a probabilistic approach, the results will be presented as distribution

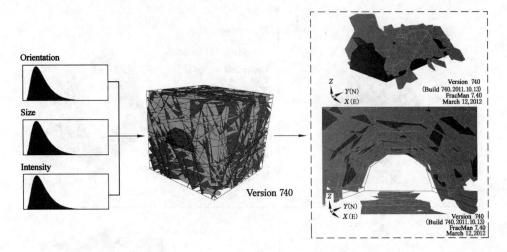

Figure 5-16 Discrete fracture network method

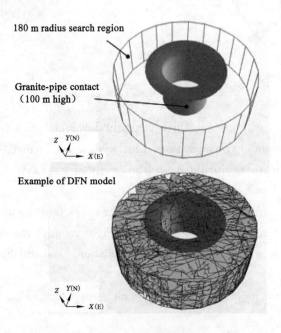

Figure 5-17 Example 3

curves. For example, percentage passing of block size and block dimensions are shown below.

Design specifications can be defined for given percentiles (e. g. 95th percentile). As shown in Figure 5-19.

(5) Example 4: fragmentation assessment:

Fragmentation is critical to the mining process. The large size of the fragmentation distribution strongly influences such issues as draw point sizing and equipment selection. The fine size of the fragmentation distribution influences the energy requirements for

Chapter 5 Rock engineering design

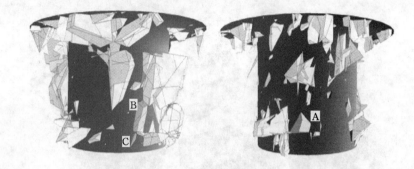

Figure 5-18 Rock bock A, B and C

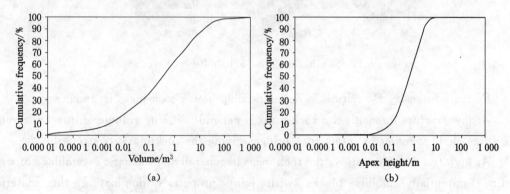

Figure 5-19 percentage passing of block size and block dimensions
(a) Kinematically unstable blocks volume;(b) Kinematically unstable blocks

crushing and milling.

Poor fragmentation assessment can be very costly.

All blocks are formed by fractures⋯but not all fractures form blocks! As shown in Figure 5-20 and Figure 5-21.

Figure 5-20 Lump rock sample

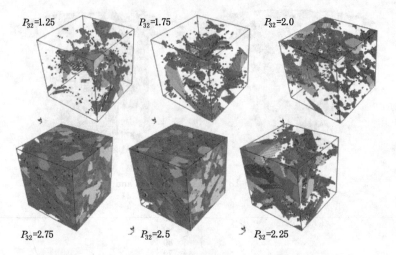

Figure 5-21   Simulation results under different P_{32} conditions

Fracture intensity P_{32} critical in understanding how a rock mass fragments.

At low fracture intensities, a rock mass is generally a large volume of intact rock and fractures, held together by intact rock bridges giving the rock mass its strength.

At high fracture intensities, the rock mass is generally a kinematic assemblage of well defined potentially mobile blocks with joint properties dominating the material strength[74,75].

The conversion from matrix to kinematic rock mass may occur over a relatively small change of P_{32}. The change diagram is shown in Figure 5-22.

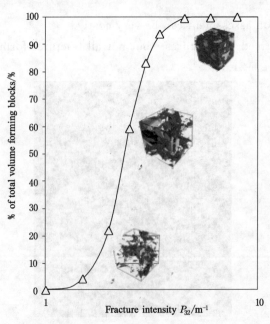

Figure 5-22   Map line changes with P_{32}

The fracture direction analysis is shown in Figure 5-23.

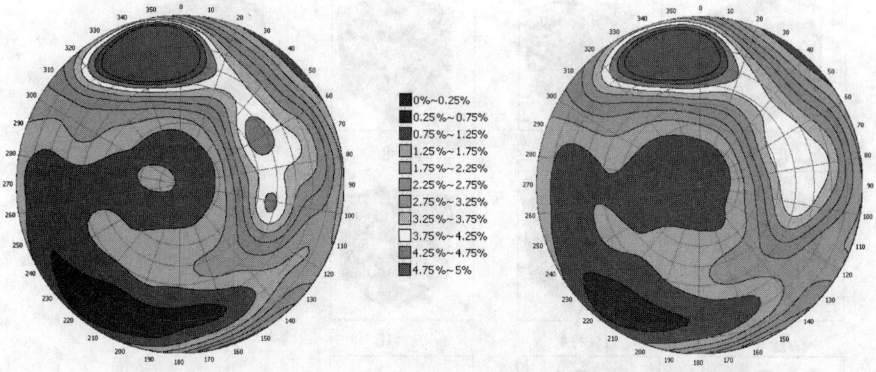

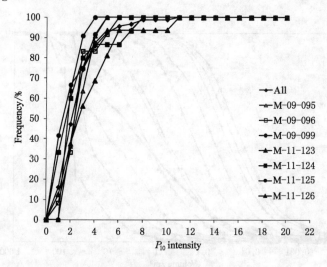

Figure 5-23 Stereonet projection of borehole data-comparison between mapped (left) and simulated (right) data

Result analysis and image display of numerical simulation (image display is shown in Figure 5-24 to Figure 5-27).

Figure 5-24 Fracture intensity analysis—P_{10} frequency curves

5.1.4 Case study: the Rio Grandee project (Hoek, 2007)

The Rio Grande pumped storage project is located on the Rio Grande river near the town of Santa Rosa de Calamucita in the Province of Cordoba in Argentina. Capacity of 1 000 MW and provides electrical storage facilities for the power grid.

Because of the excellent quality of the gneiss, most of the underground excavations did not require support.

Assessment of underground stability and installation of support, where required, was done on a 'design-as-you-go' basis which proved to be very effective and economical.

There have been no problems with rock falls or underground instability.

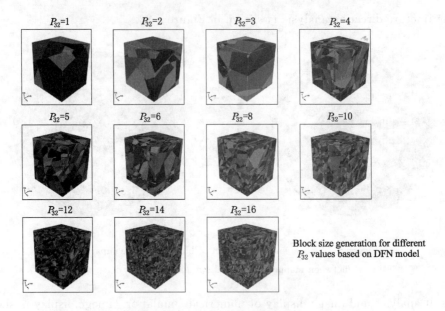

Figure 5-25　Block size generation for different P_{32} values based on DFN model

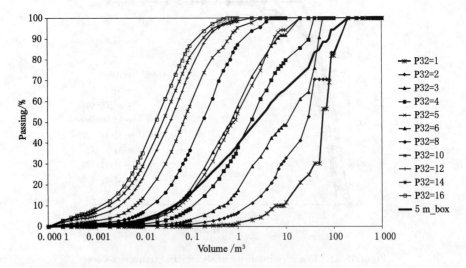

Figure 5-26　Fragmentation curves for varying fracture intensity

Water is fed directly from the reservoir down twin penstocks that then bifurcate to feed into four pump turbines, which are housed in a large underground cavern with a span of 25 m and a height of 44 m. The caverns are as shown in Figure 5-28.

Mapping of significant structural features in the roof and walls of the central access drive at the top of the cavern[76,77].

Because gneiss deposits are usually associated with tectonic deformation, projection of structural features from visible exposures would tend to be an imprecise process.

Potentially unstable blocks and wedges were reassessed after each excavation step

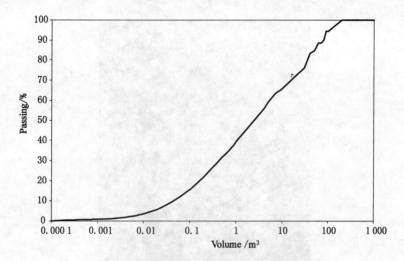

Figure 5-27 Fragmentation results (average curve)

Figure 5-28 A large underground cavern with a span of 25 m and a height of 44 m(Hoek,2007)

revealed new information.

At each stage of the cavern excavation, long rock bolts (up to 10 m length) were installed to stabilise wedges or blocks which had been determined as being potentially unstable. As shown in Figure 5-29.

Input data (UNWEDGE analysis).

Largest wedge that can occur adjacent to the excavation profile. The wedge extends over the full 25 m span of the cavern and weighs 11,610 tonnes.

Figure 5-29 Cavern

The limited extent of joints in many rock masses will restrict the size of the wedges to much smaller dimensions than those determined by using this type of analysis (UNWEDGE) for the large excavations.

Using the information about mapped trace length for joint set ♯2 (mean length=6 m) to scale the size of the rock wedges, then the resulting wedge would weigh 220 tonnes and require about seven 50 tonnes capacity fully grouted cables to give a factor of safety of about 1.5 which is considered appropriate for a cavern of this type. As shown in Figure 5-30 and Figure 5-31.

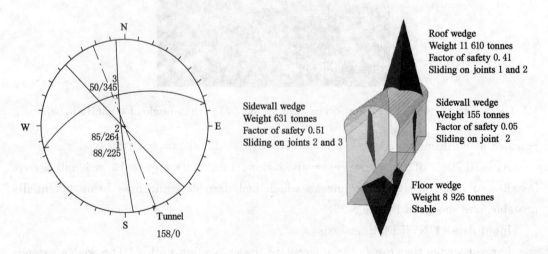

Figure 5-30 Cave model

Chapter 5　Rock engineering design

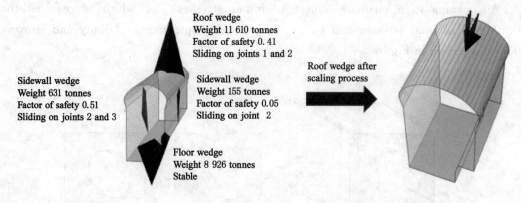

Figure 5-31　Analysis of cave model

5.2　Stress analysis & underground openings

While structurally-controlled instabilities are generally driven by gravity, stress-controlled instabilities are activated by multidirectional forces, which can be represented by a tensor with six independent components[78-80]. As shown in Figure 5-32.

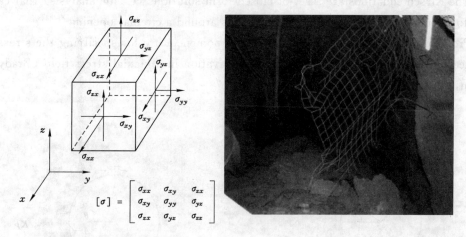

Figure 5-32　Stress analysis and chamber failure

Therefore, stress-controlled instabilities are more variable and complex.

5.2.1　Stress analysis around underground openings

Any stress analysis requires the knowledge of the magnitudes and directions of the in-situ stresses in the region of the opening.

In-situ stresses → Induced stresses.

There are several analytical solutions that can be used to calculate the induced stresses around given "simple geometries" (i.e. circular and elliptical).

Note: these solutions are given for 2-dimensional analysis.

An opening in a medium subject to initial stresses, for which is required the distribution of total stresses and excavation-induced displacements (Brady and Brown, 2004). As shown in Figure 5-33.

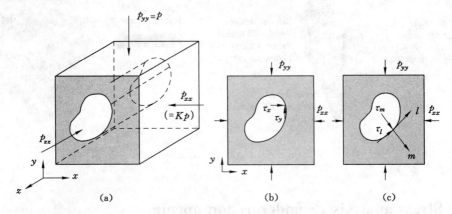

Figure 5-33 Induced stress analysis around elliptical tunnel

Numerical analysis is typically used to define the induced stresses around more complex 3-dimensional excavation geometries.

The Kirsch equations are a set of closed-form solutions (elastic analysis) that can be used to calculate the stresses and displacements around a circular opening[81,82].

Problem geometry, coordinate system and nomenclature for specifying the stress and displacement distribution around a circular excavation in a biaxial stress field (Brady and Brown, 2004). As shown in Figure 5-34.

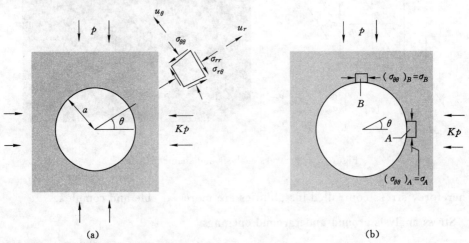

Figure 5-34 Analysis and calculation of induced stress around circular roadway

$$\sigma_{rr} = \frac{p}{2} \times \left[(1+K)\left(1 - \frac{a^2}{r^2}\right) - (1-K)\left(1 - 4 \times \frac{a^2}{r^2} + \frac{3a^4}{r^4}\right)\cos 2\theta \right]$$

$$\sigma_{\theta\theta} = \frac{p}{2} \times \left[(1+K)\left(1 + \frac{a^2}{r^2}\right) + (1-K)\left(1 + \frac{3a^4}{r^4}\right)\cos 2\theta \right]$$

$$\sigma_{r\theta} = \frac{p}{2} \times \left[(1-K)\left(1 + \frac{2a^2}{r^2} - \frac{3a^4}{r^4}\right) \sin 2\theta \right]$$

$$u_r = -\frac{pa^2}{4Gr} \times \left[(1+K) - (1-K)\left\{4(1-\nu) - \frac{a^2}{r^2}\right\} \cos 2\theta \right]$$

$$u_\theta = -\frac{pa^2}{4Gr} \times \left[(1-K)\left\{2(1-2\nu) + \frac{a^2}{r^2}\right\} \sin 2\theta \right]$$

In these expressions u_r, u_θ are displacements induced by excavation, while σ_{rr}, $\sigma_{\theta\theta}$ are total stresses after generation of the opening.

Note: "a" is the radius of the opening, "r" is the distance from the center of the opening.

By putting $r=a$ in the previous equations, the stresses on the excavation boundary are given as:

$$\sigma_{\theta\theta} = p[(1+K) + 2(1-K)\cos 2\theta]$$

$$\sigma_{rr} = 0$$

$$\sigma_{r\theta} = 0$$

The radial stresses are zero at the excavation boundary because there is no internal pressure, and the shear stress must be zero because the excavation boundary is a traction free boundary.

Similarly, for $q=0$, and r large ($\to \infty$), the stress components are given by:

$$\sigma_{\theta\theta} = 0$$

$$\sigma_{rr} = 0$$

$$\sigma_{r\theta} = 0$$

So that the far-field stresses recovered from the solutions correspond to the imposed field stresses.

5.2.2 Stress and failure criteria

The framework shown in the following slide can be used to define the suitability of a given design. Key steps in the design process are:

Determine of the stress distribution around the excavation (e.g. using the Kirsch equations).

Compare boundary stresses with the in situ crack initiation stress, σ_{ci}, and the rock mass tensile strength, σ_t or T_0.

If no boundary failure is predicted, it remains to examine the effect of any major discontinuities which will transgress the excavation[83,84].

This requires consideration of both the general effect of the structural features on boundary stresses and local stability problems in the vicinity of the discontinuity/boundary intersection.

Such considerations may lead to design changes to achieve local and more general stability conditions for the excavation perimeter. As shown in Figure 5-35.

The failure criterion is compared to the induced stresses and is not part of the

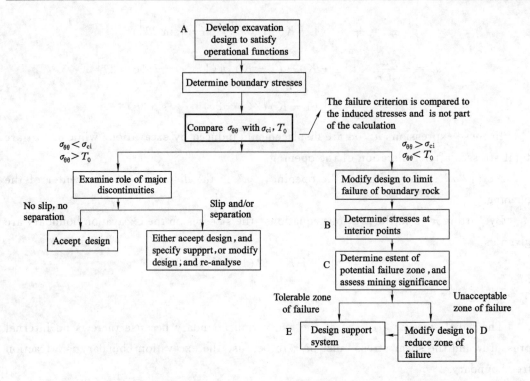

Figure 5-35   Flow chart of stress failure analysis(Brady and Brown, 2004)

calculation.

(1) Example: circular opening

A 3 m diameter tunnel is to be excavated at 450 m depth. The unit weight of the rock is 26 kN/m³, the uniaxial compressive strength and tensile strength of the of the rock are 60 MPa and 3 MPa respectively. Will the strength of the rock on the tunnel boundary be exceeded if: i) $k=0.3$ and ii) $k=2.5$? As shown in Figure 5-36.

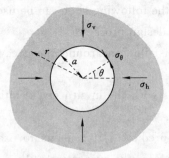

Figure 5-36   450 meters deep tunnel

You are asked to examine the stability of the rock on the boundary of the tunnel. As the tunnel has neither a support pressure nor an internal pressure applied to it, the rock on the boundary is subjected to a uniaxial state of stress[85,86], with the local $\sigma_3 = \sigma_r = 0$ and $\sigma_1 = \sigma_\theta$. At the boundary $r=a$, therefore the Kirsch Equation for tangential stress simplifies to:

$$\sigma_\theta = \sigma_v[(1+k)+2(1-k)\cos 2\theta]$$

The extreme values of induced stress occur at positions aligned with the principal in-situ stresses→Assume $q=0°$ (sidewall) and $90°$ (roof).

The concept of a zone of influence is important in mine design, since it may provide considerable simplification of a design problem.

The essential idea of a zone of influence is that it defines a domain of significant disturbance of the pre-mining stress field by an excavation. It differentiates between the near field and far field of an opening[87,88].

The stress distribution around a circular hole in a hydrostatic stress field ($k=1$), of magnitude p, is given by:

$$\sigma_{rr} = p\left(1 - \frac{a^2}{r^2}\right)$$

$$\sigma_{\theta\theta} = p\left(1 + \frac{a^2}{r^2}\right)$$

$$\sigma_{r\theta} = 0$$

Let's assume that $r=5a$, then $\sigma_{\theta\theta}$ and σ_{rr} are $1.04p$ and $0.96p$ respectively…

i.e. on the surface defined by $r=5a$, the state of stress is not significantly different (within $\pm 5\%$) from the field stresses "p". As shown in Figure 5-37.

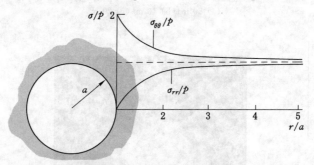

Figure 5-37 Changes in radius five times

If a second excavation (Ⅱ) were generated outside the surface described by $r=5a$ for the excavation (Ⅰ), the pre-mining stress field would not be significantly different from the in-situ stress field. As shown in Figure 5-38.

(2) Examples: zone of influence and model size

You are requested to carry out a BEM (elastic) analysis for a 10 m tunnel. How far from the excavation would you set up the outer boundary of the model? As shown in Figure 5-39 to Figure 5-43.

Since excavation Ⅰ is outside the zone of influence of excavation Ⅱ, a fair estimate of the boundary stresses around Ⅰ is obtained from the stress distribution for a single opening.

For excavation Ⅱ, the field stresses are those due to the presence of excavation Ⅰ.

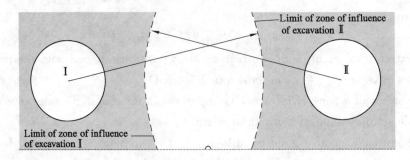

Figure 5-38　Excavation Ⅱ and excavation Ⅰ (Brady and Brown, 2004)

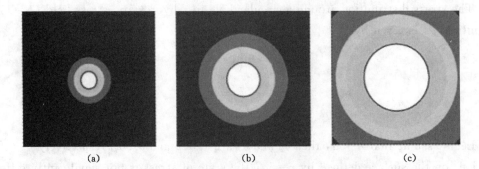

Figure 5-39　Different wide models for a 10 m tunnel
(a) 80 m; (b) 40 m; (c) 20 m

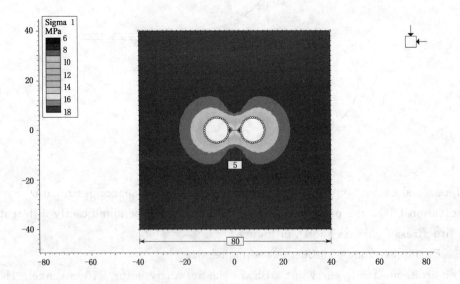

Figure 5-40　Numerical simulation image ($D=5$ m)

An engineering estimate of the boundary stresses around Ⅱ can be obtained by calculating the state of stress at the centre of Ⅱ, prior to its excavation[89,90].

This can be introduced as the far-field stresses in the Kirsch equations to yield the required boundary stresses for the smaller excavation. As shown in Figure 5-44.

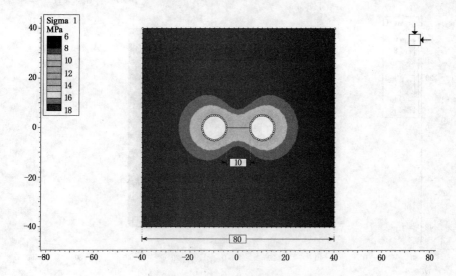

Figure 5-41　Numerical simulation image($D=10$ m)

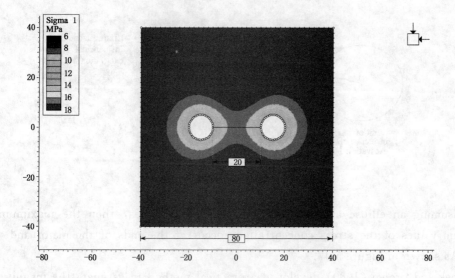

Figure 5-42　Numerical simulation image($D=20$ m)

5.2.3　Stresses around elliptical openings

The stresses around elliptical openings can be treated in an analogous way as those introduced for circular openings.

Elliptical openings can provide a first approximation to a wide variety of engineering geometries. As shown in Figure 5-45.

Assuming isotropic rock conditions, the stress state for an elliptical opening is completely characterized by two parameters:

Aspect ratio (ratio of major to minor axes).

Orientation with respect to the principal stresses. As shown in Figure 5-46.

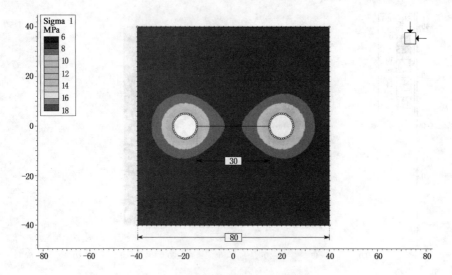

Figure 5-43   Numerical simulation image($D=30$ m)

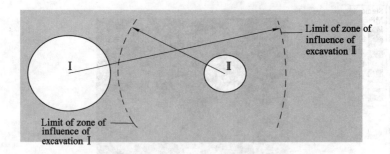

Figure 5-44   Excavation analysis

Assuming an ellipse aligned with the principal stresses, then the maximum and minimum values of the stress concentrations occur at the ends of the major and minor axes. As shown in Figure 5-47.

Hoek and Brown (1980) developed a practical method to estimate the magnitude of the tangential stresses around various types of underground openings

The method is based on a large number of detailed stress analyses undertaken using boundary element techniques.

The tangential stress in roof: $\sigma_{qr}=(Ak-1)s_v$

The tangential stress in wall: $\sigma_{qw}=(B-k)s_v$

Where A and B are roof and wall factors for various excavation shapes, k is the ratio horizontal/vertical stress, σ_v is the vertical stress. As shown in Figure 5-48.

The tangential stress in roof: $\sigma_{qr}=(Ak-1)s_v$

The tangential stress in wall: $\sigma_{qw}=(B-k)s_v$

Assume a circular tunnel, 10 m diameter, 95 m below ground level, unit weight of 0.026 MN/m³ and $k=1$.

Figure 5-45 Width/height ratio of the opening

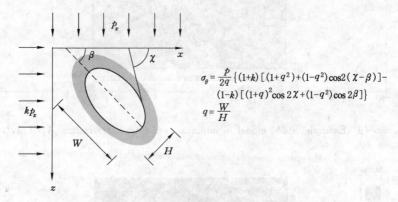

$$\sigma_\theta = \frac{p}{2q}\{(1+k)[(1+q^2)+(1-q^2)\cos 2(\chi-\beta)] - (1-k)[(1+q)^2\cos 2\chi + (1-q^2)\cos 2\beta]\}$$

$$q = \frac{W}{H}$$

Figure 5-46 Stress analysis of elliptical tunnel

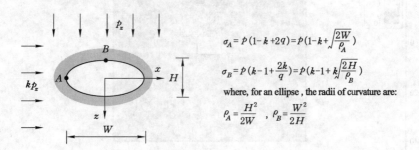

$$\sigma_A = p(1-k+2q) = p(1-k+\sqrt{\frac{2W}{\rho_A}})$$

$$\sigma_B = p(k-1+\frac{2k}{q}) = p(k-1+k\sqrt{\frac{2H}{\rho_B}})$$

where, for an ellipse, the radii of curvature are:

$$\rho_A = \frac{H^2}{2W}, \quad \rho_B = \frac{W^2}{2H}$$

Figure 5-47 An ellipse aligned with the principal stresses

$A=3$, $B=3$, $\sigma_{qr}=(3-1)2.47=4.94$ MPa and $\sigma_{qw}=(3-1)2.47=4.94$ MPa

If the tunnel is horseshoe shaped, 10 m wide:

$A=3.2, B=2.3, \sigma_{qr}=(3.2-1)2.47=5.43$ MPa and $\sigma_{qw}=(2.3-1)2.47=3.21$ MPa

The simulation results are shown in Figure 5-49 and Figure 5-50.

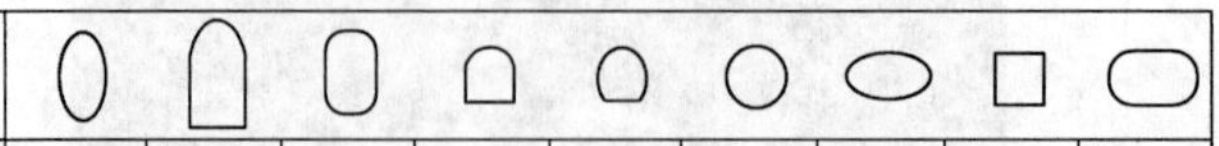

A	5.0	4.0	3.9	3.2	3.1	3.0	2.0	1.9	1.8
B	2.0	1.5	1.8	2.3	2.7	3.0	5.0	1.9	3.9

Figure 5-48　Values of constants A and B

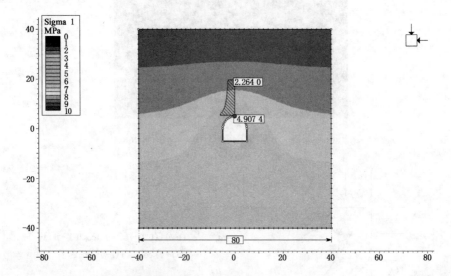

Figure 5-49　Example：BEM model & importance of grid discretization($A=3$ m,$B=3$ m)

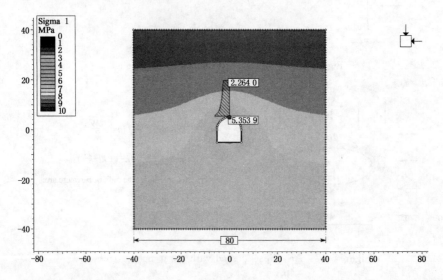

Figure 5-50　Example：BEM model & importance of grid discretization($A=3.2$ m,$B=2.3$ m)

Chapter 6 Rock mechanics design (underground)

6.1 Pillar supported mining methods

Mining methods based on pillar support are intended to control rock mass displacements throughout the zone of influence of mining, while mining proceeds.

Stope local stability and near-field ground control could be considered as separate design issues (see section 6.2) if stopes are excavated to be locally self-supporting.

Performance of a pillar support system can be expected to be related to both the dimensions of the individual pillars and their geometric location in the orebody.

Load capacity of pillars (pillar strength) vs. loads imposed on the pillars by the interacting rock mass (pillar stress). As shown in Figure 6-1.

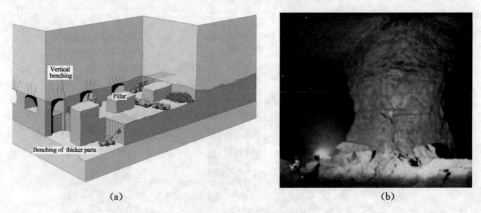

Figure 6-1 Pillar supported

Types of pillars:

Room pillar: standard pillar, analogous to a column in structural design.

Rib pillar: analogous to a continuous beam in structural design, with a width ≪ length.

Sill pillar: separates levels in a mine. Tend to be permanent, with confinement effect provided by horizontal stresses.

Crown pillar: separates underground mines from surface. Common in operations entailing interactions between open pit and block cave mining/underground mining. As shown in Figure 6-2.

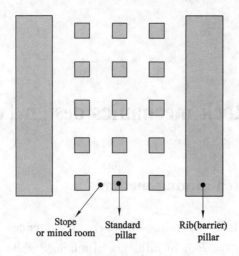

Figure 6-2 Types of pillars

6.1.1 Engineering aspects for the design of hard-rock pillars

Rock pillars can be defined as the in-situ rock between two or more underground openings (→room-and-pillar or stope-and-pillar systems).

Stoping activity in an orebody causes stress redistribution and an increase in pillar loading. As shown in Figure 6-3.

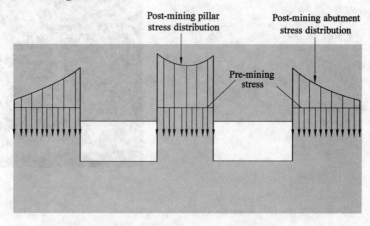

Figure 6-3 Stress distribution in rock pillar mining

For states of stress in a pillar less than the in situ rock mass strength, the pillar remains intact and responds elastically to the increased state of stress. Mining interest is usually concentrated on the peak load-bearing capacity of a pillar.

Subsequent interest may then focus on the post-peak, or ultimate load-displacement behaviour, of the pillar.

Several forms of deformation of rock pillars are shown in Figure 6-4.

Several common rock pillars are shown in Figure 6-5.

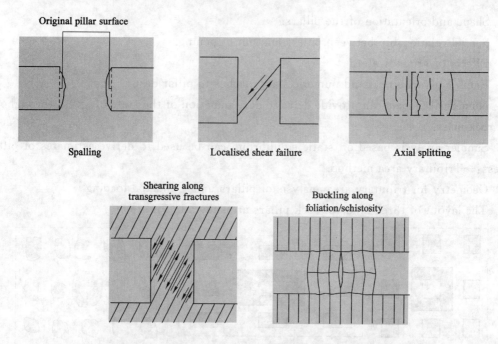

Figure 6-4 Deformations behaviours of pillars(Brady and Brown, 2004)

Figure 6-5 Pillars (examples)

6.1.2 Pillar stress (tributary area method)

The actual pillar stress is dependent on (but not limited to):

In situ stress conditions.

Mining induced stress changes.

Effects of geological structures, such as faults and jointing.

Shape and orientation of the pillars.

Spatial relationship between pillars and mine opening.

Effects of groundwater.

Empirical, analytical and numerical approaches to pillar design.

Numerical models can provide detailed determination of the state of stress throughout the rock mass.

Simpler analysis based on static equilibrium can be used to derive estimates of pillar stress: →Tributary area method.

Geometry for tributary area analysis of pillars under uniaxial loading

The layout of three kinds of rock pillars are shown in Figure 6-6.

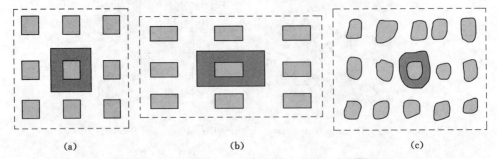

Figure 6-6 Square lay out rectangular lay out and example of as-built lay out

(1) Tributary area method(square layout)

Size and force analysis of square layout are shown in Figure 6-7.

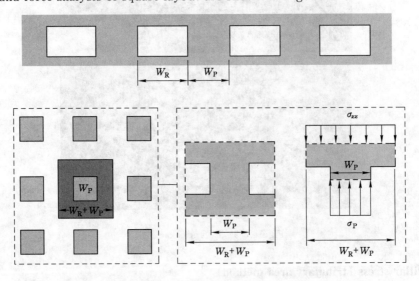

Figure 6-7 Size and force analysis of square layout

(Average) Pillar stress:

$$\sigma_P W_P^2 = \sigma_{zz} (W_R + W_P)^2$$

$$\sigma_{zz} = \gamma_R z$$

$$\sigma_P = \frac{\gamma_R z \, (W_R + W_P)^2}{W_P^2}$$

Area extraction ratio (for 2D sections):

$$R = \frac{W_R}{(W_R + W_P)}$$

$$\sigma_P = \sigma_{zz} \times \frac{1}{1-R} = \gamma_R z \times \frac{1}{1-R}$$

(2) Tributary area method(rectangular layout)(as shown in Figure 6-8)

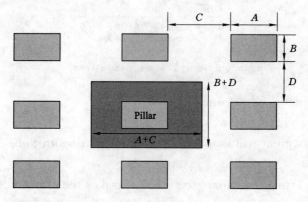

Figure 6-8 Size and force analysis of rectangular lay out

(Average) Pillar stress:

$$\sigma_P (AB) = \sigma_{zz}(A+C)(B+D)$$

$$R = \frac{[(A+C)(B+D) - AB]}{(A+C)(B+D)}$$

$$\sigma_P = \gamma_R z \times \frac{(A+C)(B+D)}{AB}$$

6.1.3 Tributary area method-observations

Pillar stress is calculated directly from the stope and pillar dimensions, and the pre-mining normal stress parallel to the pillar axis.

High incremental changes in pillar stress for small changes of the area extraction ratio.

Operational implications: e.g. extraction ratios $\geqslant 0.75$ are rare→$\sigma_P/\sigma_{zz} > 4$.

Only the normal stress component directed parallel to the principal axis of the pillar is considered[91].

Homogeneous and isotropic rock mass conditions.

Location of the pillar within the orebody or the mine is ignored→abutments effects are ignored. As shown in Figure 6-9.

6.1.4 Pillar strength

(1) The determination of pillar strength is highly dependent on the factors that affect rock mass strength:

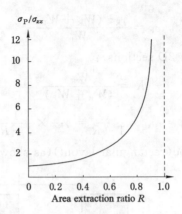

Figure 6-9 Curve of stress and ratio change

Intact rock strength and deformability.

Number, type and strength of structural features.

Pillar geometry (pillar width, pillar height and pillar width-to-height ratio $W:H$).

Blast damage.

Techniques for estimating pillar strength, defined as the ultimate load per unit area of a pillar, generally use empirical formulae based on survey data from actual mining conditions.

Note: empirical methods may fail to consider specific failure mechanisms and limitations exist associated with their intrinsic derivation from specific material properties (size, shape and stress conditions).

Empirical design methods use empirical formulae to estimate pillar strength. Pillar formulae can be subdivided in two groups:

Shape effect formulae: they assume that, for a given rock type, a pillar of a given shape, expressed in terms of its width-to-height ratio, will have a constant strength independent of changes in its size.

Size effect formulae: they are based on the principle that, for a given rock type, a pillar of given shape (pillar width to height ratio) will have reduced strength as its size increases, since samples of increasing size are thought to contain more structural discontinuities.

The Hoek-Brown failure criterion could also be used to estimate pillar strength. However, numerical simulations (continuum analysis) of pillar behaviour (hard rock) using the Hoek-Brown failure criterion should include Hoek-Brown brittle parameters to account for cohesion loss processes (paper by Martin & Maybee, 2000).

Empirical pillar formulae can be expressed according to the following general expression:

$$\sigma_P = K\left(A + B\frac{W^\alpha}{H^\beta}\right)$$

Where σ_P is the pillar strength; K represents the strength of a unit cube of the rock mass; W and H are the pillar width and pillar height respectively; The constants A, B, α and β are derived empirically.

Shape effect formulae use empirical constants α and β that are equal, meaning that pillar strength is independent of pillar volume.

Size effect formulae use empirical constants α and β that are not equal, meaning that pillar strength will decrease as pillar volume increases (for pillars of the same shape)[92].

(2) Linear shape effect formula:

The linear shape effect formula assumes that pillars with equal $W : H$ ratio will have equal strength, according to the linear relationship:

$$\sigma_P = K\left(A + B\frac{W}{H}\right)$$

Obert & Duval (1967) formula:

$$\sigma_P = K\left(0.778 + 0.222\frac{W}{H}\right)$$

This formula does not include a term to account for effects of size on strength.

Where K is the unconfined strength of a cubical pillar specimen (measured in MPa).

(3) Examples: linear shape effect formula & stability charts:

Sjoberg (1992) studied sill pillars in the Zinkgruvan Mine in Sweden ($\sigma_{ci}=240$ MPa), and found that $K=74$ MPa could be used to fit field data.

Krauland & Soder (1987) studied the pillars in the Black Angel Mine in Greenland and found that $K=35.4$ MPa could be used to fit field data (σ_{ci} up to 100 MPa). As shown in Figure 6-10.

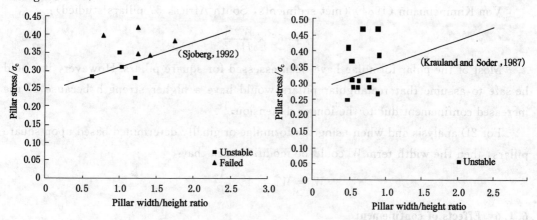

Figure 6-10 Fitting curve for effect of size

Hudyma (1988) introduced the "pillar stability graph" below for determining the strength of open stop rib pillars based on data from Canadian hard rock underground mining[93].

The valid range of pillar $W : H$ ratios for this method is [0.5, 1.4]. As shown in Figure 6-11.

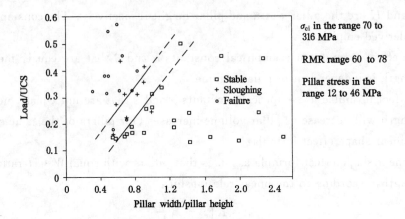

Figure 6-11 Fitting curve($W : H = [0.5, 1.4]$)

6.1.5 Size effect formula

Size effect formulae assumes that as pillars size increases, the strength of equal shape (i.e. same $W : H$ ratio) pillars will decrease:

$$\sigma_P = K \frac{W^\alpha}{H^\beta}$$

Hedley & Grant (1972)(quartzite rock type at Elliot Lake Mine, Canada. 28 pillars studied, of which 23 stable, 2 unstable and 3 failed pillars):

$$\sigma_P = 133 \times \frac{W^{0.5}}{H^{0.75}}$$

Von Kimmelmann (1984) (met sediments, South Africa, 57 pillars studied):

$$\sigma_P = 65 \times \frac{W^{0.46}}{H^{0.66}}$$

Most of the pillar formulae have been assessed for square pillars. However, it would be safe to assume that rectangular pillars would have a higher strength because of the increased confinement due to the longer dimension[94].

For 2D analysis and when using the formulae originally determined based upon square pillars, then the width term W could be modified such that:

$$W = W_e = 4 \times \frac{A_P}{P}$$

6.1.6 Effects of confinement

In all cases, the pillar-strength formulae ignore the effect of s_3 and rely on a stress to strength ratio based on the maximum pillar stress and the uniaxial compressive strength.

Lunder & Pakalnis (1997), examined the stress distribution in hard-rock pillars in Canadian mines and proposed that the average confinement in a pillar could be expressed in terms of $W : H$:

$$C_{pav} = 0.46 \times \left[\log\left(\frac{W}{H} + 0.75\right) \right]^{\frac{1.4H}{W}}$$

C_{pav} is average minor/major stress ratio within the pillar core.

However, numerical models have shown that pillar confinement depends on the ratio (k) between the far-field horizontal stress s_1 and s.

For pillar $W : H$ ratios less than 1 the effect of k can be ignored, but for $W : H > 1$, then the effect of k is significant.

Increase in confinement at the center of the pillar as a function of k, the ratio of the far-field maximum horizontal stress (σ_1) and vertical stress (σ_3).

The predicted effect of confinement (C_{pav}) using Lunder & Pakalnis (1997) is also shown (as shown in Figure 6-12).

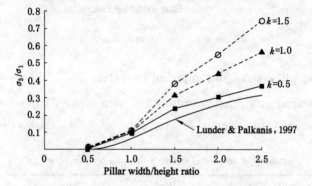

Figure 6-12 Stress and ratio variation diagram

Lunder & Pakalnis (1997) introduced the confinement formula, which combines rock mechanics with empirical methods. As shown in Figure 6-13 and Figure 6-14.

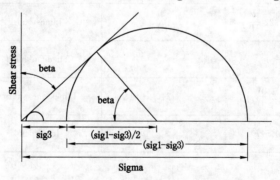

Figure 6-13 Confinement formula: mine pillar friction (k)

$$\sigma_p = 0.44 \times UCS \times (0.68 + 0.52k)$$

The confinement formula introduces a mine pillar friction (k) term calculated from the average minor/major stress ratio within the pillar core:

$$\frac{\sigma_3}{\sigma_1} = C_{pav} = 0.46 \times \left[\lg\left(\frac{W}{H} + 0.75\right) \right]^{\frac{1.4H}{W}}$$

$$k = \tan\left[\arccos\left(\frac{1 - C_{pav}}{1 + C_{pav}} \right) \right]$$

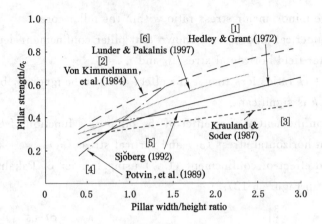

Figure 6-14　Stress ratio diagram line

6.1.7　Summary of pillar strength (empirical formulae)

Summary of empirical strength formula for hard-rock pillars where the pillar width and height is in meters (after Martin & Maybee, 2000). As shown in Table 6-1.

Table 6-1　　Summary of empirical strength formula for hard-rock pillars

	Pillar strength formulas/MPa	σ_c/MPa	Rock mass	No. of pillars
[1]	$133 \dfrac{W^{-0.5}}{H^{0.75}}$	230	Quartzites	28
[2]	$65 \dfrac{W^{-0.46}}{H^{0.66}}$	94	Metasediments	57
[3]	$35.4 \left(0.778 + 0.222 \dfrac{W}{H}\right)$	100	Limestone	14
[4]	$0.42 \sigma_c \dfrac{W}{H}$		Canadian shield	23
[5]	$74 \left(0.778 + 0.222 \dfrac{W}{H}\right)$	240	Limestone/skarn	9
[6]	$0.44 \sigma_c (0.68 + 0.52 k)$		Hard rocks	178

Doe Run lead mines in Missouri (USA).

Hedley & Grant empirical formula used to establish pillar strength, numerical analysis used to establish pillar load.

Empirical classification of pillar conditions by Roberts et al. (1998).

Specific empirical strength formulae have been derived for coal pillars:

Bieniawski (1975), with reference to large scale tests of coal specimens (Pillar width up to 2 m):

$$\sigma_p = K \left(0.64 + 0.34 \dfrac{W}{H}\right)$$

K = strength of a 30 cm³ coal specimen.

Salamon and Munro (1967):

$$\sigma_p = K \frac{W^{0.46}}{H^{0.66}}$$

$K = 7\,176$ kPa.

Sheorey (1987):

$$\sigma_p = 0.27\,\sigma_c \frac{\sqrt{W}}{H^{0.86}} \quad (W:H < 4)$$

σ_c = USC of a 25 mm³ coal speciemn

(1) Mine stiffness-energy & stability

So far it has been assumed that the pillar will fail when the average pillar stress is greater than the strength of the pillar (→FoS<1).

There is the need to better define the term failure. Some pillars may fail with some considerable violence, while others may gradually develop cracks and spall into an hour glass shape. As shown in Figure 6-15.

(a)　　　　　　　　(b)　　　　　　　　(c)

Figure 6-15　Failure of pillar

In order to understand the difference between types of pillar failure it is necessary to consider the local mine stiffness and its influence upon the behavior of the pillar.

A mine pillar could be treated as a deformable element in a soft loading system, represented by the country rock[95].

The country rock represents the spring in the loading system and the pillar represents the rock specimen. As shown in Figure 6-16.

(2) Intact rock strength

The progressive failure of the pillar depends on the relative stiffness of the pillar and the local mine stiffness.

In this case $W_m > W_s$.

Catastrophic failure occurs at, or shortly after, the peak.

The energy released by the host rock during unloading is greater than that which can

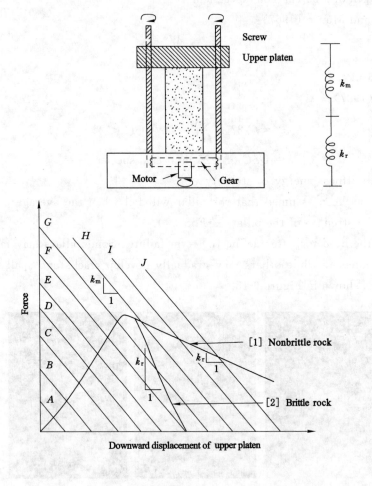

Figure 6-16　Pillar mechanics model

be absorbed by the pillar in following the post-peak curve from A to B. As shown in Figure 6-17.

In this case $W_s > W_m$.

Energy in excess of that released by the host rock as stored strain energy must be supplied in order to deform the pillar along AB.

Note that the behaviour observed up to, and including, the peak, is not influenced by the local mine stiffness. As shown in Figure 6-18.

6.1.8　Numerical modelling of pillars

Empirical methods do not consider specific failure mechanisms and their limitations are associated with their intrinsic derivation from specific material properties (observed size, shape and stress conditions).

Numerical methods overcome some of the limits of the empirical methods.

And provide an opportunity to increase our fundamental understanding of the factors

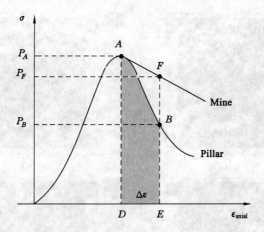

Figure 6-17 Pillar peak stress diagram

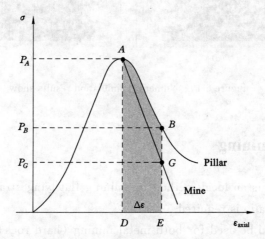

Figure 6-18 Pillar peak stress diagram

governing the strength and deformational response of jointed rock pillars.

Continuum and discontinuum models can be utilised to simulate multifracturing phenomena and the mechanical behavior of discrete systems[96].

However, neither approach alone can capture the interaction of existing discontinuities and the creation of new fractures through fracturing of the intact rock material.

Hybrid models FEM/DEM models, DEM with voronoi, or PFC models have been used increasingly in rock engineering.

Results of a series of numerical analyses of limestone pillars.

Combining the above methods in order to eliminate undesirable characteristics while retaining as many advantages as possible. As shown in Figure 6-19.

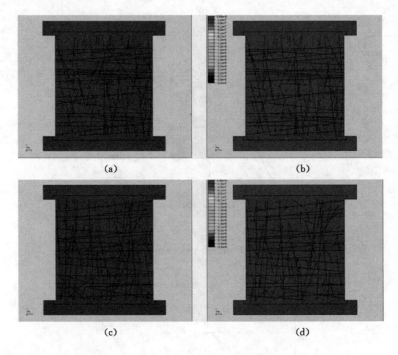

Figure 6-19   Numerical simulation results show

6.2   Longwall mining

Longwall mining is an ideal method for mining flat-lying stratiform orebody when a high area extraction ratio is required.

The method could be used for both metal mining (hard rock) and coal mining (soft rock).

The method preserves continuous behaviour of the far-field rock mass.

Longwall mining requires a reasonably uniform orebody with a dip of less than 20 degrees.

The method clearly falls between the extremes of fully supported and complete caving methods of mining.

6.2.1   Longwall coal mining(as shown in Figure 6-20)

For longwall coal mining, the immediate roof rock for the coal seam would ideally consist of relatively weak shales, siltstones or similar lithologies, with sufficient jointing to promote easy caving.

Rock quite different properties are required of the main roof, which must be sufficiently competent to bridge the span between the mine face and the consolidating bed of caved roof[97,98].

The seam floor rock must have sufficient bearing capacity to support the loads applied

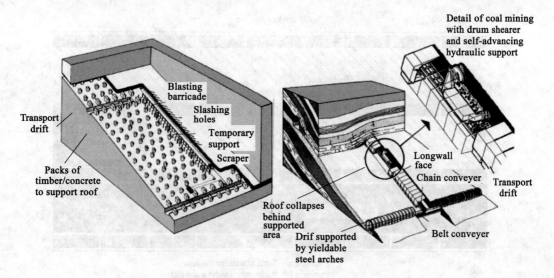

Figure 6-20 Longwall metal mining (hard rock) and longwall coal mining (soft rock) (Brady and Brown, 2004)

by the roof support system at the face line.

The uniaxial compressive strengths of the roof, seam and floor rocks are typically in the range 20~40 MPa.

The pre-mining stress could be approximately hydrostatic for weak sedimentary rocks.

A number of parallel panels or longwalls are extracted side-by-side, and the most common form of longwall coal mining is a retreat method in which the roadways are developed to the end of the panel, which is then extracted on retreat.

Pillars are left between adjacent longwalls to:

Protect the roadways from excessive displacements that may occur because of the extraction of adjacent panels.

Isolate a particular panel where unfavorable geological structures or fire, water, or gas hazards exists. Assist in controlling surface subsidence. As shown in Figure 6-21 and Figure 6-22.

6.2.2 Longwall caving mechanics

(1) Longwall caving mechanics (Brady and Brown, 2004)

Significant redistribution of the in situ stresses occurs during and following the mining of a longwall panel.

Pre-existing compressive stresses are relieved above the mined-out area and there is a concentration of compressive stresses and the development of principal stress differences in the rock over and beyond the abutments of the panel.

This causes induced tensile fracturing, de-lamination and opening of pre-existing fractures producing caving of the rock mass above the mined-out area.

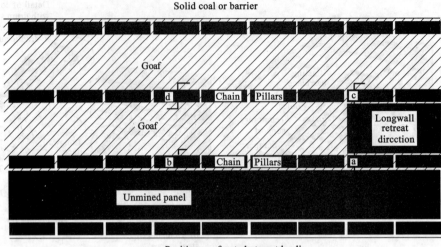

Figure 6-21　Development and extraction of adjacent longwalls(Brady and Brown, 2004)

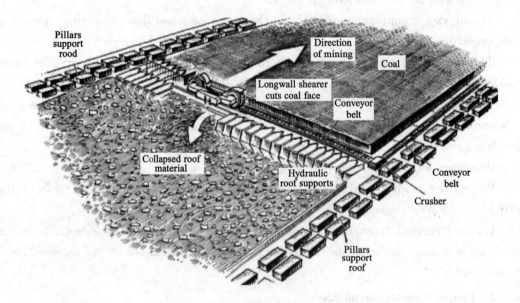

Figure 6-22　Extraction of longwalls

And shear fracturing and slip on bedding planes and natural and mining-induced discontinuities and fractures in the rock mass surrounding the panel. As shown in Figure 6-23.

(2) Roadways mechanics (Brady and Brown, 2004)

The main roadways required to service a retreating longwall are fully developed on the solid coal or unmined side of the panel before the coal is mined on retreat.

Roadways and the pillars protecting them are subjected to changing loading conditions

Chapter 6 Rock mechanics design (underground)

Figure 6-23 Fully mechanized top coal caving technology

and complex stress paths throughout their operational lives.

The roadways can be several hundreds of metres long and are required to remain serviceable for up to two years.

Roadways are generally mined to a rectangular profile (e. g. Angus Place Olliery, Western Coalfield, New South Wales, Australia, dimensions are 4.5 m wide and 3.1 m high).

(3) Roadways support (Brady and Brown, 2004)

Example of roadway support and reinforcement (Angus Place Colliery, Western Coal Field, New South Wales, Australia). 2.1 m long, grouted, high tensile steel bolts which are applied through mesh and steel straps. 4 to 8 roof bolts may be used in each row. Spacing of rows of bolts in the order of 1 to 2 m.

Cable bolts may also be used at intersections and where shorter rock bolts are inadequate. As shown in Figure 6-24.

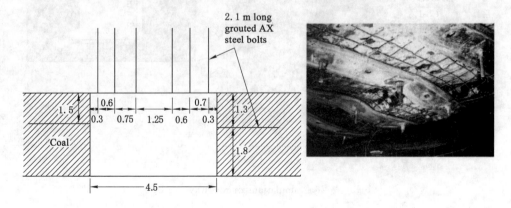

Figure 6-24 Roadway bolting(Figures after Hebblewhite, 2001)

6.2.3 Longwall caving mechanics (Fuqiang, 2012)

Longwall panel entries are usually subjected to: side abutment stress caused by the mining of adjacent panels; and front abutment pressure caused by the mining of the current panel.

The loading process is not static but dynamic as the panel is progressively excavated.

The support system is such that the roadway may not collapse. However, it is likely to "squeeze". As shown in Figure 6-25 and Figure 6-26.

Figure 6-25 Extrusion deformation on both sides of roadway

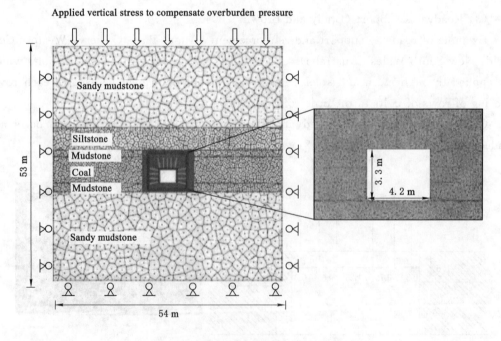

Figure 6-26 Simulations of roadway squeezing

Because this is a 2D model, induced stresses are modelled changing the boundary conditions.

In-situ stresses induced lateral abutment stresses due to mining 2202 panel (increase horizontal and vertical stresses).

Induced high front abutment pressure due to approaching of 2203 panel (increase horizontal and vertical stresses). As shown in Figure 6-27 and Figure 6-28.

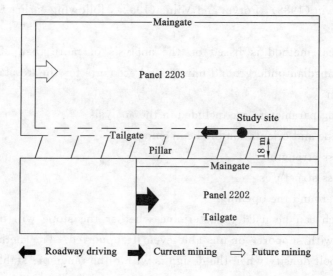

Figure 6-27　Plane plan of coal face

Figure 6-28　Numerical simulation results

6.3 Stope stability & the stability graph method

The stability graph method for open stope stability and cable bolt design was developed by Potvin (1988), Potvin and Milne (1992), following earlier work by Mathews et al. (1981).

This empirical method is based on the analysis of more than 350 case histories collected from Canadian underground mines, and accounts for the key factors influencing open stope design.

The following parameters are included in the analysis:
① Size, shape and orientation of the opening.
② Rock mass strength.
③ Rock mass structures.
④ Stresses around the opening.

The approach can be used to determine whether the stope will be stable without support, stable with support, or unstable even if supported. The method also suggests ranges of cable bolt density when the design is within the range of 'stable with support'.

6.3.1 The stability graph method

The design procedure is based upon the calculation of two factors:
① N': modified stability number which represents the ability of the rock mass to stand up under a given stress condition.
② S: shape factor or hydraulic radius which accounts for the stope size and shape.

The stability number N' is defined as:
$$N' = Q'ABC$$

Where Q' is the modified Q tunneling quality index; A is the rock stress factor; B is the joint orientation adjustment factor; C is the gravity adjustment factor.

The shape factor S is calculated as follows:
$$S = \frac{\text{Cross sectional area of surface analysed}}{\text{Perimeter of surface analysed}} \text{(Hydraulic radius)}$$

Note: Q' is calculated from the results of structural mapping of the rock mass in exactly the same way as the standard Q, except that the stress reduction factor SRF is set to 1. Since the approach has not been applied in conditions with significant groundwater, the joint water reduction factor J_w is set to 1.

The design face could be the hanging wall or the stope back.
$$HR = \frac{Wh}{2W + 2h}$$

6.3.2 Rock stress factor

(1) Rock stress factor A

The rock stress factor A reflects the stresses acting on the free surfaces of an open

Chapter 6 Rock mechanics design (underground)

stope at depth.

This factor is determined from the unconfined compressive strength of the intact rock and the stress acting parallel to the exposed face of the stope under consideration.

The induced compressive stress is generally found from numerical modelling (e. g. Examine 2D).

The rock stress factor A is determined from the σ_c/σ_1 ratio, such that (as shown in Figure 6-29):

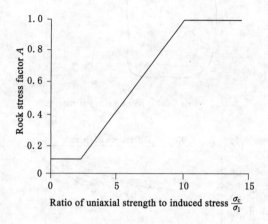

Figure 6-29 Relationship between stress ratio and stress factor A

$$\frac{\sigma_c}{\sigma_1} < 2A = 0.1$$

$$2 < \frac{\sigma_c}{\sigma_1} < 10A = 0.1125 \times \left(\frac{\sigma_c}{\sigma_1}\right) - 0.125$$

$$\frac{\sigma_c}{\sigma_1} > 10A = 10$$

(2) Joint orientation adjustment factor B

The joint orientation adjustment factor B accounts for the influence of joints on the stability of the stope faces.

Structurally controlled failure would generally occur along critical joints forming a shallow angle with the free surface. The shallower the angle between the discontinuity and the surface, the easier would be for the intact rock bridge to be broken by blasting, stress or by another joint set.

Note: a slight strength increase would occur when the angle q approaches zero, since the jointed rock blocks would act as a beam. As shown in Figure 6-30 and Figure 6-31.

(3) Adjustment for the effects of gravity C

The factor C presents an adjustment for the effects of gravity.

Failure can occur from the roof by gravity induced falls or, from the stope walls, by slabbing or sliding.

Potvin (1988) suggested that both gravity induced failure and slabbing failure could

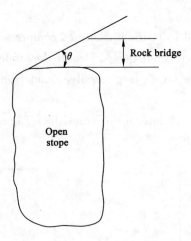

Figure 6-30 Joint azimuth map

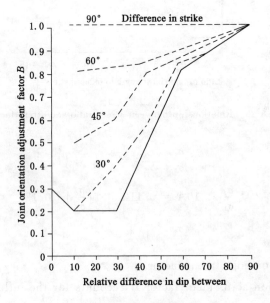

Figure 6-31 The relationship between different azimuth
and adjustment factor B

be calculated as a function of the inclination of the stope surface (α):

$$C = 8 - 6 \times \cos \alpha$$

According to the equation above, the factor C has a maximum value of 8 for vertical walls and a minimum value of 2 for horizontal stope backs. As shown in Figure 6-32.

(4) Gravity adjustment factor C for sliding failure modes

Two types of dilution:

Planned (accepted) dilution is integrated into mine-plan.

Unplanned is due to spalling or caving of stopes (unwanted tones of waste material).

Concept of ELOS (equivalent linear over-break slough).

Chapter 6 Rock mechanics design (underground)

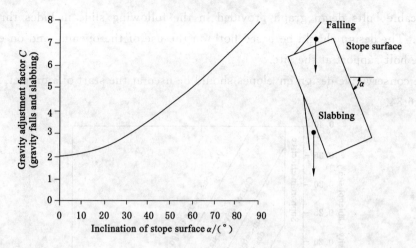

Figure 6-32 Gravity adjustment factor C for gravity falls and slabbing

It is assumed that caving within a stope can be estimated on a per m length basis, and the failure zone around the stope would have an elliptical section.

The ELOS would be the equivalent rectangular area corresponding to the caved zone in situ. As shown in Figure 6-33.

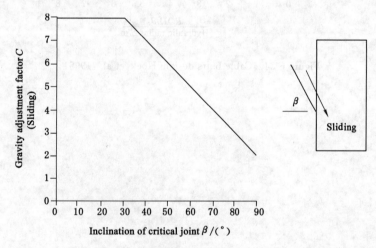

Figure 6-33 Gravity adjustment factor C for sliding failure modes

(5) Cable bolts design

When the stability analysis indicates that the stope requires support, then It is possible to provide preliminary guidelines for cable density by using an empirical correlation between frequency of jointing through the block size (parameters RQD/J_n) and the hydraulic radius of the opening.

According to Potvin et al. (1989), cable bolts are not likely to be effective when $(RQD/J_n)/HR$ is less than 0.75 and when the cable bolt density is less than 1 bolt per 10 square metres at the opening boundary.

The cable bolts deign graph provided in the following slide includes three design envelopes. The design should be based both on the use of the opening and on experience with cable bolt support at the site.

More conservative design envelopes should be used at the start of a project. As shown in Figure 6-34.

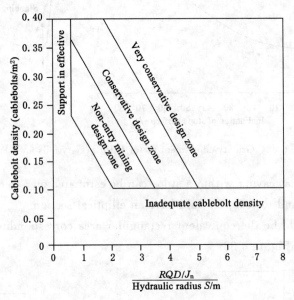

Figure 6-34　Cable bolts design(Hoek,et al.,1995)

Chapter 7 Numerical analysis

Numerical simulation, also known as numerical analysis method, is a computer program to solve the approximate solution of mathematical model, also known as computer simulation. The application of numerical simulation technology in solidification temperature field calculation has benefited from the rapid development of computer technology since 1960s. By using computer simulation technology, not only the dynamic change of temperature field is successfully solved and visually expressed, but also the theoretical basis and calculation ideas are provided for the study of other quality problems related to thermal process. For example, quantitative analysis of a series of chemical and physical metallurgical reaction processes, such as evaluation of solidification structure, solidification defects, prediction of thermal stress and residual deformation, et al. The most commonly used numerical solutions of temperature field are difference method and finite element method.

7.1 Introduction to numerical modelling

Numerical modelling plays an important role in the design of structures in rock.

The excavation of any structure in rock will ultimately result in a modification of the stress state, and an analysis of the stresses and associated displacements is required in order to understand the behavior of such excavations[99,100]. As shown in Figure 7-1.

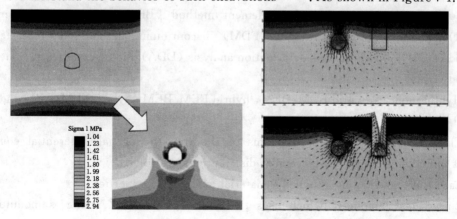

Figure 7-1 Numerical simulation analysis

The scope of the model is not to represent such processes in their entirety, rather the objective of the analyst is to determine which process need to be consideredexplicittly and which can be represented in an average way.

Remember: "we build models because the real world is too complex for our understanding; it does not help if we build models that are also too complex".

The key components of any satisfactory modelling approach are data and understanding:

A model should try and simplify reality rather than trying to create a perfect imitation.

The construction of the model should be driven by the questions that the model is supposed to answer rather than by the details of the system that is being modelled.

Simple models may be more appropriate to analyses different aspects of the problem or address the same questions from a different perspectives[101].

7.1.1 Numerical modelling

Numerical methods of stress and deformation analysis fall into two categories:

Integral methods: only the problem boundary is defined and discretized. Computationally efficient but mostly restricted to elastic analyses.

Differential methods: problem domain is defined and discretized. Non-linear and heterogeneous material properties can be included, but generally longer solution run times are required.

7.1.2 Continuum and discontinuum methods

Numerical models divide the problem (i.e. the rock mass) into zones, and...

Some zones of the rock mass could be treated as continuous, whilst discontinuum analysis would explicitly apply to other elements like discontinuities.

A continuum model would reflect mainly material deformation of the system, whilst a discontinuum model would reflect the component movement of the system.

Continuum methods: boundary element method (BEM), finite element method (FEM) and finite difference method (FDM). discontinuum methods: discrete elements method (DEM), discontinuous deformation analysis (DDA) and discrete fracture network method (DFN).

Hybrid models: hybrid BEM/DEM, hybrid FEM/BEM, hybrid FEM/DEM and other hybrid models.

The methods above differ in terms of the way the partial differential equations (PDEs) for the stress and displacement distributions are solved.

Each zone can then be assigned a material model and properties.

Linear elastic models, which uses the elastic properties (Young's modulus and Poisson's ratio) of the material.

Elastic-plastic models, which use strength parameters to limit the shear stress that a zone may sustain. As shown in Figure 7-2 and Figure 7-3.

Chapter 7 Numerical analysis

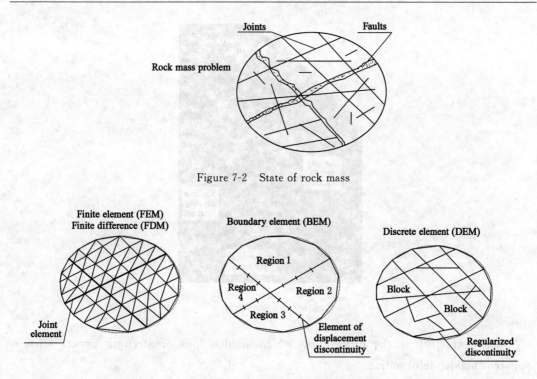

Figure 7-2 State of rock mass

Figure 7-3 Comparison of several numerical simulation methods

On a qualitative basis, the rock mass can be classified into three groups: (i) intactrock mass, (ii) fractured rock mass and (iii) highly fractured or weathered rock mass.

The mechanical behavior of (i) can be investigated using a continuum approach, whilst discontinuous model may be used for analyzing type (ii) and type (iii) rock masses (limitations with the number of details that a discontinuous model can effectively handle).

Assuming that a highly fractured rock mass behaves like a continuous body in a global sense, type (iii) rock masses could also be treated as a continuum, with equivalent material properties reflecting the effects of pre-existing fractures. As shown in Figure 7-4.

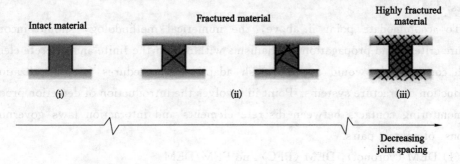

Figure 7-4 Three different fracture models of rock mass

Modelling rock brittle fracture(as shown in Figure 7-5).

Figure 7-5 Rock fracture

Brittle fracture is the process by which sudden loss of strength occurs with no apparent plastic deformation.

Models can be used to better understand the process through which rock material fails and to "predict" failure.

7.1.3 Continuous/discontinuous transformations

In a modelling perspective, brittle failure corresponds to a transition from a continuum to a discontinuum state. Key factors in the numerical approach include:

i. Constitutive models that govern the material failure.

ii. Ability of the numerical approach to introduce discontinuities such as shear bands and cracks generated during the material failure and fracture process.

iii. Effective simulation of contact between the region boundaries and crack surfaces during the failure process and particle behavior motion of fragments in post-failure phases[102,103].

To accommodate point ii above, the numerical methodology should incorporate fracture criteria and propagation mechanisms within both the finite and discrete elements, which consequently would require mesh adaptively procedures for discretization and introduction of fracture systems. Point iii involves the introduction of detection procedures for monitoring contacts between discrete elements and interaction laws governing the response of contact pairs.

(1) DEM (voronoi), DEM (PFC) and FEM/DEM

The DEM (voronoi) involves generation of a large number of random, irregular and interconnected polygonal blocks. The boundaries of these blocks are treated as contacts in the DEM code. The behavior of those connected blocks is highly dependent on the assumed properties for the block boundaries.

DEM particle flow code (PFC), in which brittle fracture is simulated as the result of bond breakage between particles.

Hybrid continuum-discontinuum techniques (FEM/DEM), in which brittle fracture is simulated using an adaptive meshing scheme associated with a specific constitutive criterion. As shown in Figure 7-6.

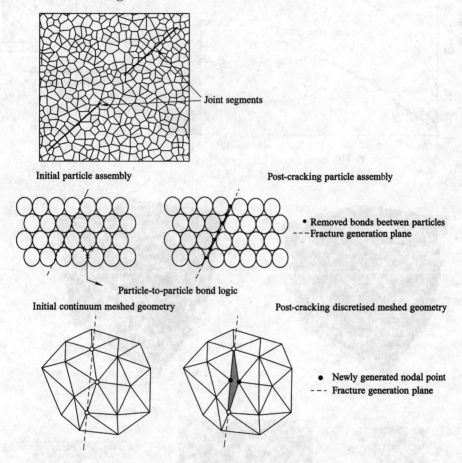

Figure 7-6   DEM (voronoi), DEM (PFC) and FEM/DEM

(2) Synthetic rock mass (SRM) modelling

Both 2D and 3D DEM techniques have been applied to the development of synthetic rock mass (SRM) properties.

The objective is to combine the effects of the intact and fractured portions of the rock mass into a unique set of equivalent continuum properties[104].

This approach allows engineers to model equivalent Mohr-Coulomb or Hoek-Brown strength envelopes, including anisotropic effects, by running suitable biaxial (2D) and triaxial (3D) test models of fractured rock masses. As shown in Figure 7-7.

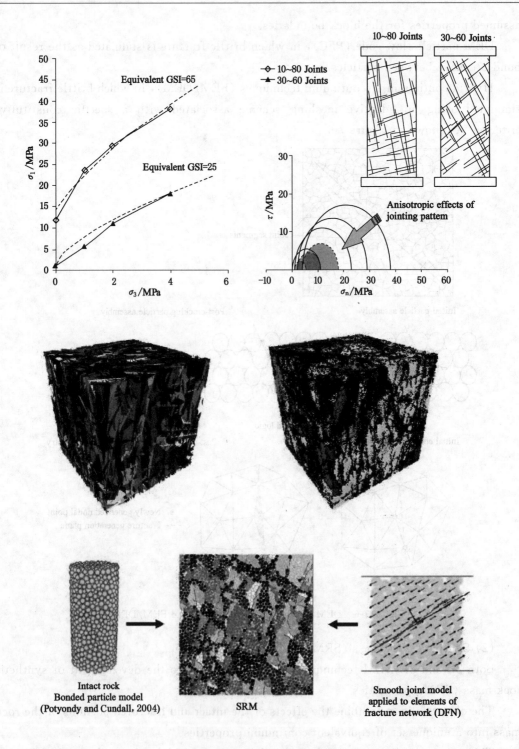

Figure 7-7 Synthetic rock mass (SRM) modelling

7.2 Applications of numerical models to coal mining

7.2.1 Numerical modelling of rock slopes

Continuum modelling is typically used for the analysis of rock masses made of massive, intact rock, weak rocks and soil like heavily jointed rock masses.

Discontinuum modelling is more appropriate for rock masses controlled by discontinuity behavior.

Hybrid codes (e.g. FDEM) can be used to couple continuum and discontinuum techniques. As shown in Figure 7-8.

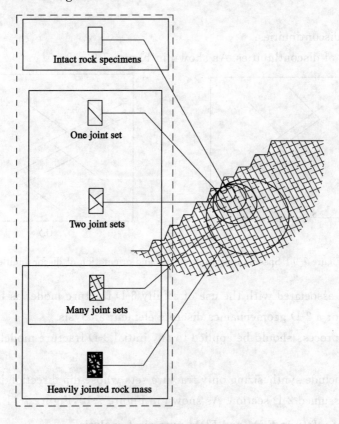

Figure 7-8　Numerical modelling of rock slopes

Numerical modelling: know its place…and limitations:

Modelling requires that the real problem be idealized, or simplified, in order to fit the constraints imposed by factors such as available material models and computer capacity.

Analysis of rock mass response involves different scales. It is impossible (and undesirable) to include all features, and details of rock mass response mechanisms, into one model.

Remember that many of the details of rock mass behavior are unknown and unknowable; therefore, the approach to modelling is not as straightforward as it is, say, in other branches of mechanics[105].

There are known knowns. These are things we know that we know. There are known unknowns.

That is to say, there are things that we know we don't know. But there are also unknown unknowns. There are things we don't know we don't know (Donald Rumsfeld).

7.2.2 Structural input data & numerical models

When using a discontinuum model (but this limitations also apply to continuum models) caution should be taken that the structural input into the analysis is representative.

Spacing of discontinuities.

Orientation of discontinuities. As shown in Figure 7-9.

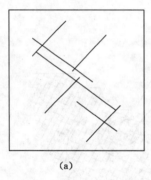 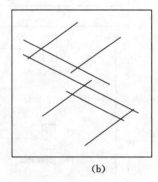

(a)   (b)

Figure 7-9   Spacing of discontinuities and orientation of discontinuities

Limitations associated with the use of a fully 3-D fracture model as the source of 2-D fracture traces for a 2-D geomechanics discrete element analysis.

A filtering process should be applied to any initial 3-D fracture model prior to use in a 2-D analysis.

This may include synthesizing only fracture sets whose dip direction is parallel to subparallel to the assumed 2-D section. As shown in Figure 7-10, Figure 7-11 and Figure 7-12.

7.2.3 Factor of safety in FEM and FDM numerical analysis

For numerical analysis of slopes, we can define the factor of safety as the ratio of the actual shear strength to the minimum shear strength required to prevent failure.

Within a FEM or FDM model the FoS can be computed by reducing the shear strength until collapse occurs. This shear-strength reduction technique was used first with finite elements by Zienkiewicz et al. (1975)[106,107].

To perform slope stability analysis with the shear strength reduction technique, simulations are run for a series of increasing trial factors of safety (f). Shear strength properties, cohesion (C) and friction angle (φ), are reduced for each trial according to the

Chapter 7　Numerical analysis

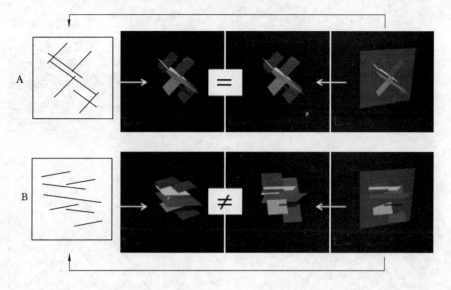

Figure 7-10　Discrete element numerical analysis

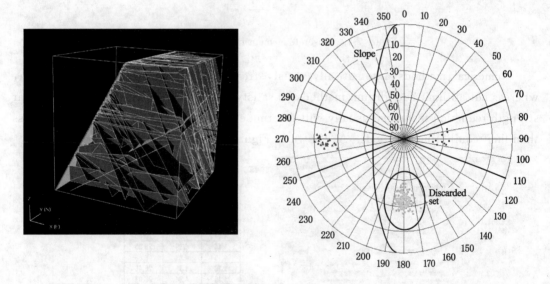

Figure 7-11　Discrete element numerical analysis model

equations.

$$C_{\text{trial}} = \left(\frac{1}{f}\right)C$$

$$\varphi_{\text{trial}} = \arctan\left(\frac{\tan\varphi}{f}\right)$$

7.2.4　Shear-strength reduction analysis

Two main advantages over limit equilibrium slope stability analyses:

The critical slide surface is found automatically, and it is not necessary to specify the shape of the slide surface (e.g. circular, log spiral, piecewise linear) in advance.

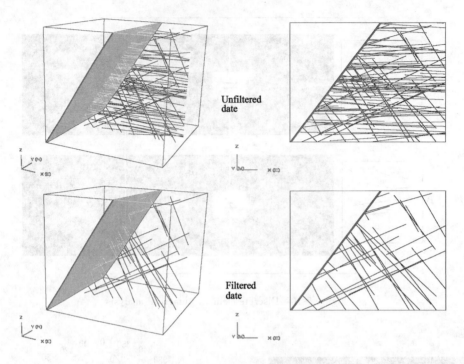

Figure 7-12 Discrete element numerical analysis

Numerical methods automatically satisfy translational and rotational equilibrium, whereas not all limit equilibrium methods do satisfy equilibrium. Consequently, the shear strength reduction technique usually will determine a safety factor equal to or slightly less than limit equilibrium methods. As shown in Figure 7-13.

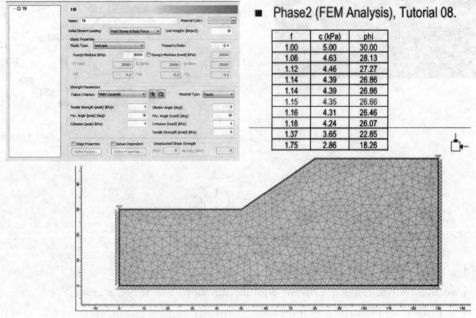

Figure 7-13 Example: shear-strength reduction analysis

7.3 Discrete fracture network modelling

7.3.1 DFN modelling

(1) What is discrete fracture network (DFN) modelling

The DFN approach seeks to explicitly represent discrete fractures in 3D Space.

Build a model of key fracture geometries, including mechanical and hydraulic properties.

Derive the input properties from DFN analysis of available geotechnical data.

Results in the generation of geologically realistic DFN models based on and calibrated back to parameters derived from field data. As shown in Figure 7-14 and Figure 7-15.

Figure 7-14 DFN three dimensional discrete model

It is a stochastic process allowing multiple but equi-probable realisations to be created.

(2) DFN modelling today

The DFN modelling approach has been used on numerous projects in civil, mining, and oil and gas projects where a good understanding of the effects of fracture networks is required.

The technique has advanced from the simple fractures in a box problem to more complex geocellular and stratigraphic models. As shown in Figure 7-16.

(3) DFN modelling requires certain fracture properties to be defined

As shown in Figure 7-17.

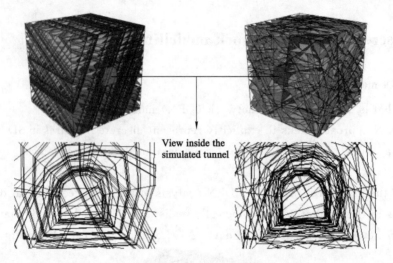

Figure 7-15   DFN three dimensional discrete model

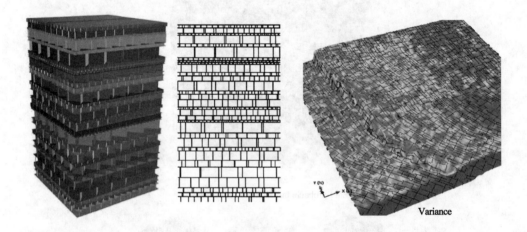

Figure 7-16   DFN modelling

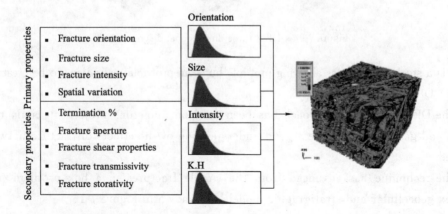

Figure 7-17   Fracture properties to be defined for DFN modelling

(4) Spatial models

The starting point for the generation of a DFN model is the definition of the spatial model that governs the way fractures are generated within a given 3D volume.

The majority of the spatial models involve similar considerations for specific fracture characteristics, such as shape (generally polygons), size and termination at intersections. The main differences lie in the specific distribution laws used to simulate fracture orientation and fracture location.

Examples of fracture spatial models include a) Enhanced Baecher, b) Nearest-Neighbor and c) Fractal Levy-Lee. As shown in Figure 7-18.

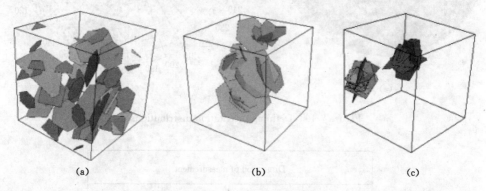

Figure 7-18 DFN 3D model generated

(5) Defining orientation distributions

Conventional orientation analysis concentrates upon the main clusters of orientation data rather than the whole distribution. This can result in as little as 50% of the data being categorized[108].

DFN based orientation analysis seeks to fully define 100% of the data into their appropriate sets based upon a range of differing orientation distributions.

When the data are highly dispersed, using a bootstrapping technique may be more appropriate.

This is a statistical method based upon multiple random sampling with replacement from an original sample to create a pseudo-replicate sample of fracture orientations.

Basically, it uses field data to produce a similar but slightly different fracture population. As shown in Figure 7-19.

Typically, in the DFN community fracture intensity is expressed with reference to a unified system of fracture intensity measures that provide an easy framework to move between differing scales and dimensions.

Fracture intensity is referred to as P_{ij} intensity, where the subscript i refers to the dimensions of sample, and the subscript j refers to the dimensions of measurement.

For example, volumetric fracture intensity (P_{32}) is defined as the ratio of total fracture area to unit volume (dimensions of m^2/m^3). As shown in Figure 7-20.

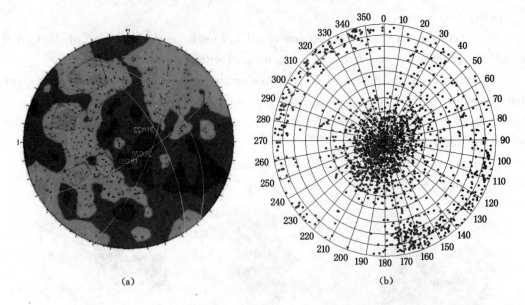

Figure 7-19 Defining orientation distributions

		Dimension of measurement				
		0	1	2	3	
Dimension of sample	1	P_{10} No of fractures per unit length of borehole	P_{11} Length of fractures per unit length			Linear measures
	2	P_{20} No of fractures per unit area	P_{21} Length of fractures per unit area	P_{22} Area of fractures per area		Areal measures
	3	P_{30} No of fractures per unit volume		P_{32} Area of fractures per unit volume	P_{33} Volume of fractures per unit volume	Volumetric measures
		Density		Intensity	Porosity	

Figure 7-20 Fracture intensity definitions

P_{32} is an intrinsic rock mass property and whilst it cannot be directly measured, it can be inferred from either 1D or 2D data using a simulated sampling methodology on the basis of a linear correlation with P_{21} (m/m^2), which is the total trace length of fractures per unit area, or with P_{10} (m^{-1}), which represents the total number of fractures along a scanline or borehole (i.e. P_{10} is a measure of fracture frequency):

$$P_{32} = C_{32} P_{21}$$

$$P_{32} = C_{31} P_{10}$$

7.3.2 Deriving fracture intensity (P_{32} correlation with P_{10})

Determine P_{32} by simulation:

Take orientation and size distribution data. Simulate a model with an initial P_{32} value.

Sample the model in the same way as your data (e.g. borehole or trace plane) and derive P_{10} or P_{21} data.

Repeat for a number of P_{32} values.

Specify P_{10} directly.

Set the P_{10} value for a simulated borehole (or a number of boreholes) and generate fractures until the target P_{10} value is reached. As shown in Figure 7-21.

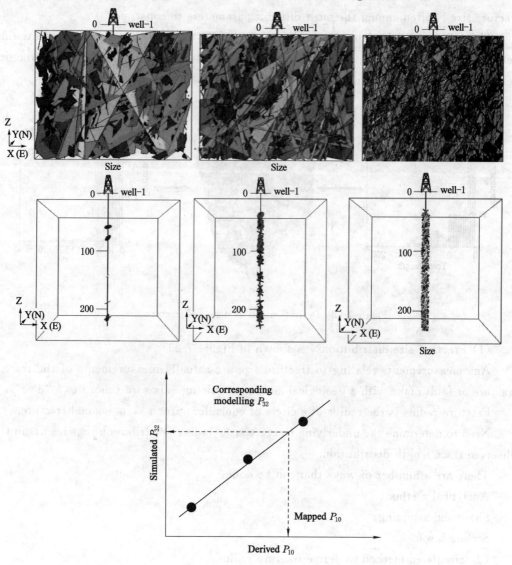

Figure 7-21 Deriving fracture intensity (P_{32} correlation with P_{10})

7.3.3 Deriving fracture intensity (P_{32} correlation with P_{10})

For each fracture set:

Select spatial model, orientation distribution and fracture radius distribution. Use an estimated P_{32} intensity value (P_{32} sim).

Generate DFN model and sample P_{21} on plane corresponding to mapped rock exposure Repeat process for three estimates of P_{32} sim, then estimate linear relationship between P_{32} sim and P_{21} sim.

Use mapped P_{21} to calculate P_{32} for field conditions.

7.3.4 Fracture size

The derivation of fracture size distribution is critical to any DFN modelling; however fracture size is often among the most difficult parameters to constrain.

The definition of fracture size in the context of DFN modelling requires differentiating between fracture trace length (mapped on rock exposures, also referred to as fracture persistence) and fracture radius. As shown in Figure 7-22.

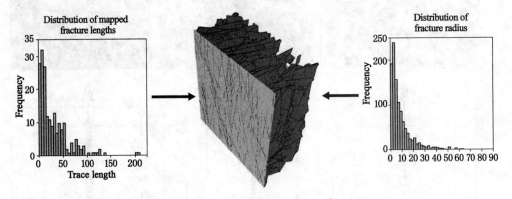

Figure 7-22   Derivation and modeling of fracture size

(1) Fracture size distributions (as shown in Figure 7-23)

Any measurements relating to fracture size are actually measurements of the trace a fracture or fault make with a geological surface or mining exposure (chord to a "disc").

Fracture radius is the radius of a circle of equivalent area to a polygonal fracture.

Need to determine the underlying fracture size (radius) distribution that results in the observed trace length distribution.

There are a number of ways that can be done:

Analytical method;

Simulated sampling;

Scaling laws.

(2) Simulated method to derive fracture radius

Trace mapping data from drifts/walls can be used to derive the underlying fracture

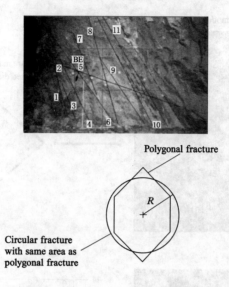

Figure 7-23 Fracture size distributions

radius distribution.

Generate trace intersections on an plane equivalent to the drifts/walls an determine length distribution.

Compare the results to the mapped trace length distribution and iterated until a best fit distribution is achieved. As shown in Figure 7-24.

(3) DFN model validation

When generating large scale DFN models, a significant degree of model validation is required as the modelling volume, and its internal variability, it is much greater than for small scale DFN models.

The main spatially varying properties that need to be validated within the large scale DFN models are overall fracture intensity and fracture orientation.

DFN models can be constrained by using a range of data sources/types such as:

1D data: borehole/scan line data (used for defining fracture orientation, intensity, aperture, geotech domains).

2D data: face, bench, outcrop mapping, photogrammetry (used for defining orientation, intensity, termination percentage, length scale, geotech domains).

3D data: geocellular input from structural restoration, 3D seismic data (e.g. velocity, coherency), curvature analysis, et al.

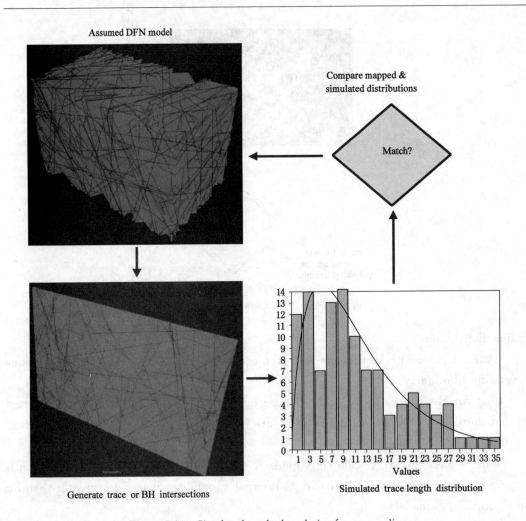

Figure 7-24　Simulated method to derive fracture radius

References

[1] Akram M S, Zeeshan M. Rock Mass Characterization and Support Assessment along Power Tunnel of Hydropower in Kohistan Area, KPK, Pakistan[J]. Journal of the Geological Society of India,2018,91(2):221-226.

[2] Peng R, Meng X, Zhao G, et al. Experimental research on the structural instability mechanism and the effect of multi-echelon support of deep roadways in a kilometre-deep well[J]. Plos One,2018,13(2):0192470.

[3] Silva J C, Milestone N B. Cement/rock interaction in geothermal wells. The effect of silica addition to the cement and the impact of CO_2, enriched brine[J]. Geothermics,2018,73(7):16-31.

[4] Wu G, Chen W, Tian H, et al. Numerical evaluation of a yielding tunnel lining support system used in limiting large deformation in squeezing rock[J]. Environmental Earth Sciences,2018,77(12):439.

[5] Palmstrøm A. Characterizing rock masses by the RMi for use in practical rock engineering: Part 1: The development of the Rock Mass index (RMi)[J]. Tunnelling & Underground Space Technology,1996,11(2):175-188.

[6] Yang S Q, Jing H W, Li Y H, et al. An Experimental Study of the Fracture Coalescence Behaviour of Brittle Sandstone Specimens Containing Three Fissures[J]. Rock Mechanics & Rock Engineering,2012,45(4):563-582.

[7] Wang Q Z, Gou X P, Zhang S. Recalibration and Clarification of the Formula Applied to the ISRM-Suggested CCNBD Specimens for Testing Rock Fracture Toughness[J]. Rock Mechanics & Rock Engineering,2013,46(2):303-313.

[8] Bieniawski Z T. 22-Classification of Rock Masses for Engineering: The RMR System and Future Trends[J]. Rock Testing & Site Characterization,1993:553-573.

[9] Ozturk C A, Nasuf E. Strength classification of rock material based on textural properties[J]. Tunnelling & Underground Space Technology Incorporating Trenchless Technology Research,2013,37(6):45-54.

[10] Zhang Q Z, Jang H S, Bae D S, et al. Empirical rock mechanical site-descriptive modeling (RMSDM) for the Korea Atomic Energy Research Institute Underground Research Tunnel (KURT)[J]. Environmental Earth Sciences,2016,75(10):860.

[11] Sonmez H, Gokceoglu C, Ulusay R. An application of fuzzy sets to the Geological Strength Index (GSI) system used in rock engineering[J]. Engineering Applications of Artificial Intelligence,2003,16(3):251-269.

[12] Sonmez H, Gokceoglu C, Ulusay R. Indirect determination of the modulus of deformation of rock masses based on the GSI system[J]. International Journal of Rock Mechanics & Mining Sciences,2004,41(5):849-857.

[13] Osgoui R R,Ünal E. An empirical method for design of grouted bolts in rock tunnels based on the Geological Strength Index (GSI)[J]. Engineering Geology,2009,107(3):154-166.

[14] Sheng-Ming H U, Xiu-Wen H U. Estimation of rock mass parameters based on quantitative GSI system and Hoek-Brown criterion[J]. Rock & Soil Mechanics,2011,32(3):861-866.

[15] Haftani M, Chehreh H A, Mehinrad A, et al. Practical Investigations on Use of Weighted Joint Density to Decrease the Limitations of RQD Measurements[J]. Rock Mechanics & Rock Engineering,2016,49(4):1551-1558.

[16] Zhang L. Determination and applications of rock quality designation (RQD)[J]. Journal of Rock Mechanics & Geotechnical Engineering,2016,8(3):389-397.

[17] Zhang L. Determination and applications of rock quality designation (RQD)[J]. Journal of Rock Mechanics & Geotechnical Engineering,2016,8(3):389-397.

[18] Labuz J F. Mohr-Coulomb Failure Criterion [J]. Rock Mechanics & Rock Engineering,2012,45(6):975-979.

[19] Labuz J F, Zang A. Mohr-Coulomb Failure Criterion[C]//The ISRM Suggested Methods for Rock Characterization, Testing and Monitoring: 2007-2014. Springer International Publishing,2015.

[20] Rahman M A, Anand S C. Empirical Mohr-Coulomb Failure Criterion for Concrete Block-Mortar Joints[J]. Journal of Structural Engineering,1994,120(8):2408-2422.

[21] Pirani F, Cappelletti D, Liuti G. Range, strength and anisotropy of intermolecular forces in atom-molecule systems: an atom-bond pairwise additivity approach[J]. Chemical Physics Letters,2001,350(3):286-296.

[22] Kirchner H P, Gruver R M. Strength-Anisotropy-Grain Size Relations in Ceramic Oxides[J]. Journal of the American Ceramic Society,2010,53(5):232-236.

[23] Huang Z G, Lai H, Zhang J M, et al. The Influences of Size and Anisotropy Strength on Hysteresis Scaling for Anisotropy Heisenberg Multilayer Films[J]. Solid State Phenomena,2007(121-123):1085-1088.

[24] Ma L J, Fang Q, Xu H F, et al. A New Elasto-Viscoplastic Damage Model Combined with the Generalized Hoek-Brown Failure Criterion for Bedded Rock Salt and its Application[J]. Rock Mechanics & Rock Engineering,2013,46(1):53-66.

[25] Yuan-Yao L I, Yin K L, Dai Y X. Stability analysis of rock slope by strength reduction method based on generalized Hoek-Brown failure criterion[J]. Rock & Soil Mechanics,2008.

[26] Zong Q B, Wei-Ya X U. Stability analysis of excavating rock slope using generalized Hoek-Brown failure criterion[J]. Rock & Soil Mechanics,2008,29(11):3071-3076.

References

[27] Ma L J, Liu X Y, Fang Q, et al. An elasto viscoplastic damage model combined with generalized Hoek Brown failure criterion for rock salt and its engineering application [J]. Journal of China Coal Society,2012,37(37):139-141.

[28] Zhu Z, Zhang G, Zhu J, et al. Research on equivalent strength rock mass based on generalized Hoek-Brown failure criterion[J]. Coal Geology & Exploration,2012,40(4):52-55.

[29] Press C. Rock Engineering Design[J]. Crc Press,2013.

[30] Feng Xia-Ting. Rock engineering design[M]. CRC Press/Balkema,1980.

[31] Feng X T, Hudson J A. Specifying the information required for rock mechanics modelling and rock engineering design[J]. International Journal of Rock Mechanics & Mining Sciences,2010,47(2):179-194.

[32] Brown E T. Rock engineering design of post-tensioned anchors for dams—A review [J]. Journal of Rock Mechanics & Geotechnical Engineering,2015,7(1):1-13.

[33] Bieniawski Z T. Design Methodology in Rock Engineering[J]. Crc Press,1992.

[34] Kotthaus S, Smith T E L, Wooster M J, et al. Derivation of an urban materials spectral library through emittance and reflectance spectroscopy[J]. Isprs Journal of Photogrammetry & Remote Sensing,2014,94(8):194-212.

[35] Agneta F, Melissa C, Yager P L, et al. Antarctic sea ice carbon dioxide system and controls[J]. Journal of Geophysical Research Oceans,2011,116(C12):12035.

[36] Haltuchmelissa A, Puntandré E. The promises and pitfalls of including decadal-scale climate forcing of recruitment in groundfish stock assessment[J]. Canadian Journal of Fisheries & Aquatic Sciences,2011,68(5):912-926.

[37] Heist E J. Population Genetics of the Sandbar Shark (Carcharhinus plumbeus) in the Gulf of Mexico and Mid-Atlantic Bight[J]. Copeia,1995,1995(3):555.

[38] Fuchsman C A, Murray J W, Konovalov S K. Concentration and natural stable isotope profiles of nitrogen species in the Black Sea[J]. Marine Chemistry,2008,111(1):90-105.

[39] Kong D, Lu S, Ping P. Linking Safety Factor and Probability of Failure Based on Monte Carlo Simulation in Fire Safety Design[C]//Fire Science and Technology 2015. Springer Singapore,2017.

[40] Wilson D, Filion Y R, Moore I D. Identifying Factors that Influence the Factor of Safety and Probability of Failure of Large-diameter, Cast Iron Water Mains with a Mechanistic, Stochastic Model: A Case Study in the City of Hamilton[J]. Procedia Engineering,2015(119):130-138.

[41] Barbosa M R, Morris D V, Sarma S K. Factor of safety and probability of failure of rockfill embankments[J]. Géotechnique,1989,39(3):471-483.

[42] Lumb P. Safety factors and the probability distribution of soil strength[J]. Canadian Geotechnical Journal,1970,7(3):225-242.

[43] Lumb P. Safety factors and the probability distribution of soil strength[J]. Canadian

Geotechnical Journal,1970,7(3):225-242.

[44] Castillo E,Conejo A J,MÍNguez R,et al. An alternative approach for addressing the failure probability-safety factor method with sensitivity analysis[J]. Reliability Engineering & System Safety,2003,82(2):207-216.

[45] Huang B. Researches on Excavation Methods for Weak Surrounding Rock in Highway Tunnel[J]. Soil Engineering & Foundation,2004.

[46] Chandra S,Nilsen B,Lu M. Predicting excavation methods and rock support: a case study from the Himalayan region of India[J]. Bulletin of Engineering Geology & the Environment,2010,69(2):257-266.

[47] Fan Y,Lu W,Zhou Y,et al. Influence of tunneling methods on the strainburst characteristics during the excavation of deep rock masses[J]. Engineering Geology, 2016(201):85-95.

[48] Niu X,Minghui Y U,Hou G. Analysis on Influence of Excavation Methods of Large-span Double-arch Tunnel on Stability of Surrounding Rock[J]. Tunnel Construction, 2011.

[49] Fekete S,Diederichs M,Lato M. Geotechnical and operational applications for 3-dimensional laser scanning in drill and blast tunnels[J]. Tunnelling & Underground Space Technology,2010,25(5):614-628.

[50] Thuro K. Drillability prediction: geological influences in hard rock drill and blast tunnelling[J]. Geologische Rundschau,1997,86(2):426-438.

[51] Wang M. Key techniques for a shield to pass through the tunnel section excavated by drill and blast[J]. Modern Tunnelling Technology,2008.

[52] Huang D. TBM Tunneling and Drill-and Blast Method Combined Construction Techniques[J]. Underground Engineering & Tunnels,2005.

[53] Liu G F,Feng X T,Feng G L,et al. A Method for Dynamic Risk Assessment and Management of Rockbursts in Drill and Blast Tunnels[J]. Rock Mechanics & Rock Engineering,2016,49(8):1-23.

[54] Ramulu M,Chakraborty A K,Sitharam T G. Damage assessment of basaltic rock mass due to repeated blasting in a railway tunnelling project—A case study[J]. Tunnelling & Underground Space Technology,2009,24(2):208-221.

[55] Fullelove I,Onederra I,Villaescusa E. Empirical approach to estimate rock mass damage from long-hole winze (LHW) blasting[J]. Mining Technology, 2017, 126 (1):34-43.

[56] Feng S J,Guo P,Sun S G,et al. The Research of Blasting Dynamices to Slope Rock Mass Damage[J]. Advanced Materials Research,2014(962-965):384-387.

[57] Liu H Y,Wang M. Numerical Analysis of Damage to Retained Rock Mass of Dam Foundation Caused by Blasting Excavation[J]. Advanced Materials Research, 2011 (255-260):4256-4261.

[58] Kulsrestha R. Blasting Techniques to Reduce Insitu Rock Damage & Improving

Highwall Stability[J]. Pesq. agropec. bras. ,2012(50).

[59] Inyang H I, Pitt J M. The rock fracture hammer for indexing the impact strength of rocks in mechanical excavation[J]. International Journal of Rock Mechanics & Mining Sciences & Geomechanics Abstracts, 1993, 30(7):707-710.

[60] Sato T, Kikuchi T, Sugihara K. In-situ experiments on an excavation disturbed zone induced by mechanical excavation in Neogene sedimentary rock at Tono mine, central Japan[J]. Engineering Geology, 2000, 56(1-2):97-108.

[61] Omori K. Hydrochemical disturbances measured in groundwater during the construction and operation of a large-scale underground facility in deep crystalline rock in Japan[J]. Environmental Earth Sciences, 2015, 74(4):3041-3057.

[62] Zou D. Mechanical Underground Excavation in Rock[C]//Theory and Technology of Rock Excavation for Civil Engineering. Springer Singapore, 2017.

[63] Dahmen N J, Turner J. Mechanical Excavation of Roadways and Chambers in Hard Rock[C]//High Level Radioactive Waste Management 1992. ASCE, 2011.

[64] Sato T, Kikuchi T, Sugihara K. In-situ experiments on an excavation disturbed zone induced by mechanical excavation in Neogene sedimentary rock at Tono mine, central Japan[J]. Engineering Geology, 2000, 56(1-2):97-108.

[65] Inyang H I, Pitt J M. The rock fracture hammer for indexing the impact strength of rocks in mechanical excavation[J]. International Journal of Rock Mechanics & Mining Sciences & Geomechanics Abstracts, 1993, 30(7):707-710.

[66] Wang J H. Analysis on mechanism and effect of rock bolts and cables in gateroad with coal seam as roof[J]. Journal of China Coal Society, 2012, 37(1):1-7.

[67] Pelto, Elmer M. Method and apparatus for anchoring rock bolts and cables[J]. Oulsnam, 1987.

[68] Li W T, Wang Q, Li S C, et al. Deformation and failure mechanism analysis and control of deep roadway with intercalated coal seam in roof[J]. Journal of China Coal Society, 2014, 39(1):47-56.

[69] Yang D C, Gao M Z, Cheng Y H, et al. Analysis on instability of surrounding rock in gob-side entry retaining with the character of soft rock composite roof[J]. Advanced Materials Research, 2012(524-527):396-403.

[70] Li X, Aziz N, Mirzaghorbanali A, et al. Behavior of Fiber Glass Bolts, Rock Bolts and Cable Bolts in Shear[J]. Rock Mechanics & Rock Engineering, 2016, 49(7):1-13.

[71] Stille H, Holmberg M, Nord G. Support of weak rock with grouted bolts and shotcrete[J]. International Journal of Rock Mechanics & Mining Sciences & Geomechanics Abstracts, 1989, 26(1):99-113.

[72] Lee S, Kim D, Ryu J, et al. An experimental study on the durability of high performance shotcrete for permanent tunnel support[J]. Tunnelling & Underground Space Technology, 2006, 21(3):431-431.

[73] Chen J X, Jiang J C, Wang M S. Function of Rock Bolt of Lattice Girder and Shotcrete

Support Structure in Loess Tunnel [J]. China Journal of Highway & Transport,2007.

[74] Franzén T. Shotcrete for underground support: a state-of-the-art report with focus on steel-fibre reinforcement[J]. Tunnelling & Underground Space Technology,1992,7(4):383-391.

[75] Liu K Y,Qiao C S,Tian S F. Design of tunnel shotcrete-bolting support based on a support vector machine[J]. International Journal of Rock Mechanics & Mining Sciences,2004,41(3):510-511.

[76] Agliardi F,Crosta G B,Meloni F,et al. Structurally-controlled instability,damage and slope failure in a porphyry rock mass[J]. Tectonophysics,2013,605(605):34-47.

[77] Agliardi F,Crosta G B,Meloni F,et al. Kink zone localization, structurally-controlled instability, and large-scale rock slope failure at the Mt. Gorsa porphyry quarry (Trentino,Italy)[C] // EGU General Assembly Conference. EGU General Assembly Conference Abstracts,2012:11617.

[78] Taheri A. An investigation of a structurally-controlled rock cut instability at a metro station shaft in Esfahan, Iran [C]//The Tenth International Symposium on Landslides and Engineered Slopes. 2008:481-485.

[79] Liu W,Hou Z. A new approach to suppress spillover instability in structural vibration control[J]. Structural Control & Health Monitoring,2010,11(1):37-53.

[80] Michoud C,Abellan A,Baillifard F J,et al. The structurally-controlled rockslide of Barmasse (Valais, Switzerland): structural geology, ground-based monitoring and displacement vs. rainfall modeling[C]//EGU General Assembly Conference. EGU General Assembly Conference Abstracts,2012.

[81] Lu H. Rock Anisotropy and Stress Analysis of Underground Openings[J]. Chinese Journal of Rock Mechanics & Engineering,1989.

[82] Hocking G,Brown E T,Watson J O. Three dimensional elastic stress analysis of underground openings by the boundary integral equation method: third symposium on engng applications of solid mechanics, toronto, June 1976 summaries [J]. International Journal of Rock Mechanics & Mining Sciences & Geomechanics Abstracts,1976,13(9):A105-A105.

[83] Kumar P,Singh B. Thermal stress analysis of underground openings[J]. International Journal for Numerical & Analytical Methods in Geomechanics,2010,13(4):411-425.

[84] Sharan S K. Finite element analysis of underground openings[M]. 1989.

[85] Chang T Y,Zhang X L,Yang S P,et al. Stress and Fracture Analysis of Underground Openings[J]. South African Historical Journal,2015,55(2006):20-32.

[86] WANG Meng, QI Te, QIN Hongyan, et al. Numerical simulation of the reasonable size of the pillar with Wongawilli mining method[J]. Journal of Earth Environment, 2018.

[87] Brady B H G,Brown E T. Naturally supported mining methods[J]. 1999.

References

[88] Zhu G H, Ping-An H U. Optimize of Ore Pillar Mining Methods Based on Fuzzy Synthetical Judgment[J]. Hunan Nonferrous Metals, 2009.

[89] Ren Y F, Ji Q X. Study on characteristic of stress field in surrounding rocks of shallow coalface under long wall mining[J]. Journal of China Coal Society, 2011, 36(10):1612-1618.

[90] Ren Y F, Liu J, Qing-Xin Q I. Movement Characteristic of Overlying Strata Structure over Long-wall Mining Face in Shallow Buried Coal Seam [J]. Coal Mining Technology, 2011.

[91] Song W, Xu W, Du J, et al. Stability of workface using long-wall mining method in extremely thin and gently inclined iron mine[C]//International Symposium on Mine Safety Science & Engineering, 2012:624-628.

[92] Mawdesley C, Trueman R, Whiten W J. Extending the Mathews stability graph for open-stope design[J]. Transactions of the Institution of Mining & Metallurgy, 2001, 110(1):27-39.

[93] Suorineni F T, Kaiser P K, Tannant D D. Likelihood statistic for interpretation of the stability graph for open stope design[J]. International Journal of Rock Mechanics & Mining Sciences, 2001, 38(5):735-744.

[94] Brady B H G, Brown E T. Pillar supported mining methods [M]. Springer Netherlands, 2004.

[95] ZHU W C, LI, et al. Numerical simulation on rockburst of underground opening triggered by dynamic disturbance[J]. Tunnelling & Underground Space Technology Incorporating Trenchless Technology Research, 2010, 25(5):587-599.

[96] Mitri H S, Hughes R, Zhang Y. New rock stress factor for the stability graph method [J]. International Journal of Rock Mechanics & Mining Sciences, 2011, 48(1):141-145.

[97] Fidelis Tawiah Suorineni. The stability graph after three decades in use: Experiences and the way forward[J]. International Journal of Surface Mining Reclamation & Environment, 2010, 24(4):307-339.

[98] Uesaka M, Kawamata T, Kaneko N, et al. Introduction to numerical modelling[C]// Cellular Growth of Crystals. Springer Berlin Heidelberg, 1991.

[99] Morscheidt W, Cavadias S, Rousseau F, et al. Pollution of the Rhine River: An introduction to numerical modelling[J]. Education for Chemical Engineers, 2013, 8(4):119-123.

[100] Caloz C. Introduction to the special issue 'Numerical Modelling of Metamaterial Properties, Structures and Devices' [J]. International Journal of Numerical Modelling Electronic Networks Devices & Fields, 2010, 19(2):83-85.

[101] Shi B, Hou Z. Application of Engineering Geological Model in Research of Water-preserved Coal Mining[J]. Mining Research & Development, 2006.

[102] Qi L I. Numerical Simulation and Application of Coal Seam Gas Occurrence in Deep

Coal Mining[J]. Coal Technology, 2018.

[103] Zhang H W, Wang Y L, Zheng X P, et al. Application of Computer Numerical Simulation Technology on Casting Process of Coal Mining Machine Cutting Tooth Base[J]. Coal Mine Machinery, 2016.

[104] Chaoshui Xu, Peter Dowd. A new computer code for discrete fracture network modelling[J]. Computers & Geosciences, 2010, 36(3): 292-301.

[105] Tran N H, Chen Z, Rahman S S. Integrated conditional global optimisation for discrete fracture network modelling[J]. Computers & Geosciences, 2006, 32(1): 17-27.

[106] Xu C, Dowd P. A new computer code for discrete fracture network modelling[M]. Pergamon Press, Inc. , 2010.

[107] Lambert C, Giacomini A, Casagrande D, et al. Rockfall Hazard Analysis From Discrete Fracture Network Modelling with Finite Persistence Discontinuities[J]. Rock Mechanics & Rock Engineering, 2012, 45(5): 871-884.

[108] Jones M A, Pringle A B, Fulton I M, et al. Discrete fracture network modelling applied to groundwater resource exploitation in southwest Ireland[J]. Geological Society London Special Publications, 1999, 155(1): 83-103.

Appendix to section 3.1 RMR tables

Table 1 RMR (1976) table

A. Classification parameters and their ratings

	Parameter		Ranges of values				For this low range ** uniaxial compressive test is preferred		
1	Strength of intact rock material	Point load strength index	>8 MPa	4~8 MPa	2~4 MPa	1~2 MPa	10~25 MPa	3~10 MPa	1~3 MPa
		Uniaxial compressive strength	>200 MPa	100~200 MPa	50~100 MPa	25~50 MPa			
	Rating		15	12	7	4	2	1	0
2	Drill core quality RQD		90%~100%	75%~90%	50%~75%	25%~50%	<25%		
	Rating		20	17	13	8	3		
3	Spacing of joints		>3 m	1~3 m	0.3~1 m	50~300 mm	<50 mm		
	Rating		30	25	20	10	5		
4	Condition of joints		Not continuous Hard joint wall rock Very rough surfaces No separation	Continuous Hard joint wall rock Slightly rough surfaces Separation <1 mm	Continuous Soft joint wall rock Slightly rough surfaces Separation <1 mm	Continuous Gouge <5 mm thick Slickensided surfaces Joints open 1~5 mm	Continuous Soft gouge >5 mm thick Slickensided surfaces Joints open >5 mm		
	Rating		25	20	12	6	0		

Continued Table 1

		None	<25 litres/min	25~125 litres/min	>125 litres/min
Ground water	Inflow per 10 m tunnel length	None	<25 litres/min	25~125 litres/min	>125 litres/min
	Ratio: Joint water pressure / Major principal stress	0	0~0.2	0.2~0.5	>0.5
	General conditions	Completely dry	Moint only (interstitial water)	Water under moderate pressure	Severe water problems
	Rating	10	7	4	0

B. Rating adjustment for joint orientations (see E)

Strike and dip orientations of joints		Very favourable	Favourable	Fair	Unfavourable	Very unfavourable
Ratings	Tunnels	0	−2	−5	−10	−12
	Foundations	0	−2	−7	−15	−25
	Slopes	0	−5	−25	−50	−60

C. Rock mass classes determined from total ratings

Rating	100~81	80~61	60~41	40~21	<21
Class No.	I	II	III	IV	V
Description	Very good rock	Good rock	Fair rock	Poor rock	Very poor rock

D. meaning of rock mass classes

Class No.	I	II	III	IV	V
Average stand-up time	10 years for 5 m span	6 months for 4 m span	1 week for 3 m span	5 hours for 1.5 m span	10 min for 0.5 m span
Cohesion of the rock mass	>300 kPa	200~300 kPa	150~200 kPa	100~150 kPa	<100 kPa
Friction of the rock mass	>45°	40°~45°	35°~40°	30°~35°	<30°

E. The effect of joint strike and dip orientations in tunnelling

Strike perpendicular to tunnel axis				Strike parallel to tunnel axis		Dip 0°~20° irrespective of strike
Drive with dip		Drive against dip				
Dip 45°~90°	Dip 20°~45°	Dip 45°~90°	Dip 20°~45°	Dip 45°~90°	Dip 20°~45°	Dip 0°~20° irrespective of strike
V. favourable	Favourable	Fair	Unfavourable	V. unfavourable	Fair	Unfavourable

Appendix to section 3.1 RMR tables

Table 2   RMR (1976) table

A. Classification parameters and their ratings

	Parameter		Range of values						
1	Strength of intact rock material	Point-load strength index	>10 MPa	4~10 MPa	2~4 MPa	1~2 MPa	For this low range-uniaxial compressive test is preferred		
		Uniaxial comp. strength	>250 MPa	100~250 MPa	50~100 MPa	25~50 MPa	5~25 MPa	1~5 MPa	<1 MPa
		Rating	15	12	7	4	2	1	0
2	Drill core Quality RQD		90%~100%	75%~90%	50%~75%	25%~50%	<25%		
	Rating		20	17	13	8	3		
3	Spacing of discontinuities		>2 m	0.6~2 m	200~600 mm	60~200 mm	<60 mm		
	Rating		20	15	10	8	5		
4	Condition of discontinuities (See E)		Very rough surfaces Not continuous No separation Unweathered wall rock	Slightly rough surfaces Separation <1 mm Slightly weathered walls	Slightly rough surfaces Separation <1 mm Highly weathered walls	Slickensided surfaces or Gouge <5 mm thick or Separation 1~5 mm Continuous	Soft gouge >5 mm thick or Separation >5 mm Continuous		
	Rating		30	25	20	10	0		
5	Groundwater	Inflow per 10 m tunnel length (l/m)	None	<10	10~25	25~125	>125		
		(Joint water press)/(Major principal σ)	0	<0.1	0.1~0.2	0.2~0.5	>0.5		
		General conditions	Completely dry	Damp	Wet	Dripping	Flowing		
	Rating		15	10	7	4	0		

B. Rating adjustment for discontinuity orientations (See F)

Strike and dip orientations		Very favourable	Favourable	Fair	Unfavourable	Very unfavourable
Ratings	Tunnels & mines	0	−2	−5	−10	−12
	Foundations	0	−2	−7	−15	−25
	Slopes	0	−5	−25	−50	

Continued Table 2

C. Rock mass classes determined from total ratings

Rating	100~81	80~61	60~41	40~21	<21
Class number	I	II	III	IV	V
Description	Very good rock	Good rock	Fair rock	Poor rock	Very poor rock

D. Meaning of rock classes

Class number	I	II	III	IV	V
Average stand-up time	20 years for 15 m span	1 year for 10 m span	1 week for 5 m span	10 hrs for 2.5 m span	30 min for 1 m span
Cohesion of rock mass (kPa)	>400	300~400	200~300	100~200	<100
Friction angle of rock mass (deg)	>45	35~45	25~35	15~25	<15

E. Guidelines for classification of discontinuity conditions

Discontinuity length (persistence)	<1 m	1~3 m	3~10 m	10~20 m	>20 m
Rating	6	4	2	1	0
Separation (aperture)	None	<0.1 mm	0.1~1.0 mm	1~5 mm	>5 mm
Rating	6	5	4	1	0
Roughness	Very rough	Rough	Slightly rough	Smooth	Slickensided
Rating	6	5	3	1	0
Infilling (gouge)	None	Hard filling <5 mm	Hard filling >5 mm	Soft filling <5 mm	Soft filling >5 mm
Rating	6	4	2	2	0
Weathering	Unweathered	Slightly weathered	Moderately weathered	Highly weathered	Decomposed
Rating	6	5	3	1	0

F. Effect of discontinuity strike and dip orientation in tunnelling**

Strike perpendicular to tunnel axis			Strike parallel to tunnel axis	
	Drive with dip-Dip 20°~45°		Dip 45°~90°	Dip 20°~45°
Drive with dip-Dip 45°~90°	Favourable		Very unfavourable	Fair
Very favourable				
Drive against dip-Dip 45°~90°	Drive against dip-Dip 20°~45°		Dip 0°~20°-Irrespective of strike	
Fair	Unfavourable		Fair	

Some conditions are mutually exclusive. For example, if infilling is present, the roughness of the surface will be overshadowed by the influence of the gouge. In such cases use A. 4 directly. ** Modified after Wickham et al. (1972).